Studies in Advanced Mathematics

Pseudodifferential Analysis on Symmetric Cones

Studies in Advanced Mathematics

Series Editor

STEVEN G. KRANTZ
Washington University in St. Louis

Editorial Board

R. Michael Beals
Rutgers University

Dennis de Turck
University of Pennsylvania

Ronald DeVore
University of South Carolina

L. Craig Evans
University of California at Berkeley

Gerald B. Folland
University of Washington

William Helton
University of California at San Diego

Norberto Salinas
University of Kansas

Michael E. Taylor
University of North Carolina

Titles Included in the Series

Real Analysis and Foundations, *Steven G. Krantz*
Fast Fourier Transforms, *James S. Walker*
Measure Theory and Fine Properties of Functions, *L.Craig Evans and Ronald Gariepy*
Partial Differential Equations and Complex Analysis, *Steven G. Krantz*
The Cauchy Transform Theory, Potential Theory, and Conformal Mapping, *Steven R. Bell*
Several Complex Variables and the Geometry of Real Hypersurfaces, *John P. D'Angelo*
An Introduction to Operator Algebra, *Kehe Zhu*
Modern Differential Geometry of Curves and Surfaces, *Alfred Gray*
Wavelets: Mathematics and Applications, *John Benedetto and Michael W. Frazier*
A Guide to Distribution Theory and Fourier Transforms, *Robert Strichartz*
Invariance Theory, the Heat Equation, and the Atiyah-Singer Index Theorem, *Peter B. Gilkey*
Course in Abstract Harmonic Analysis, *Gerald B. Folland*
Dynamical Systems: Stability, Symbolic Dynamics, and Chaos, *Clark Robinson*
Fourier Analysis and Partial Differential Equations, *José Garcia-Cuerva, Eugenio Hernández, Fernando Soria, and José-Luis Torrea*
Practical Guide to Wavelets, *Bjorn Jawerth and W. Sweldens*
Wavelets and Other Orthgonal Systems with Applications, *Gilbert G. Walter*
Pseudodifferential Analysis on Symmetric Cones, *André Unterberger and Harald Upmeier*
Clifford Algebras in Analysis and Related Topics, *John Ryan*
Composition Operators on Space of Analytic Functions, *Carl C. Cowen, Jr., and Barbara I. MacCluer*

ANDRÉ UNTERBERGER
University of Reims, France

HARALD UPMEIER
University of Marburg, Germany

Pseudodifferential Analysis on Symmetric Cones

CRC PRESS
Boca Raton New York London Tokyo

Library of Congress Cataloging-in-Publication Data

Unterberger, André.
Pseudodifferential analysis of symmetric cones / André Unterberger, Harald Upmeier.
p. cm. — (Studies in advanced mathematics)
Includes bibliographical references (p. –) and index.
ISBN 0-8493-7873-7 (alk. paper)
1. Pseudodifferential operators. 2. Boundary value problems. 3. Cones (Operator theory) I. Upmeier, Harald, 1950– . II. Title. III. Series.
QA329.7.U585 1995
515'.7242—dc20 95-34182
CIP

Direct all inquiries to CRC Press, Inc., 2000 Corporate Blvd., N.W., Boca Raton, Florida 33431.

Internationhal Standard Book Number 0-8493-7873-7
Library of Congress Card Number 95-34182
Printed in the United States of America 1 2 3 4 5 6 7 8 9 0
Printed on acid-free paper

Preface

The analysis of boundary value problems on domains with nonsmooth boundary has become an active area of research in partial differential equations. For this study it is essential to have a good symbolic calculus of pseudodifferential operators reflecting the geometry of the boundary singularities. This book develops such a calculus for singularities modeled by arbitrary "symmetric" cones in $\mathbb{R}^n$. Symmetric cones, possibly disguised under nonlinear changes of coordinates, are the building blocks of manifolds with edges, corners or conical points of a very general nature.

Besides being a canonical open set of some euclidean space, a symmetric cone Λ has an intrinsic Riemannian structure of its own, that turns it into a symmetric space. These two structures make it possible to define on Λ a pseudodifferential analysis (the Fuchs calculus), which is, in some sense, as precise as the one available on $\mathbb{R}^n$. The methods used throughout rely heavily on the use of tools from quantum mechanics, such as representation theory and coherent states. Classes of operators defined by their symbols are given intrinsic characterizations; this makes the analysis quite neat at some points (e.g., the study of the composition of operators) where traditional methods would be much too complicated if not hopeless.

Harmonic analysis is the subject of later parts of the book, via the automorphism group of the complex tube over Λ. The holomorphic discrete series of representations of this group plays a role in the definition of coherent states; also, the Fuchs calculus may be thought of as arising from a contraction of a one-parameter family of calculi linked to this series.

This monograph is intended for use by researchers in pseudodifferential analysis. While the emphasis is mainly on analysis on nonsmooth domains, the precision (and, most of the time, novelty as well) of the methods should be of interest to anyone working in pseudodifferential analysis. Also, people interested in the geometry or harmonic analysis of symmetric cones will find here a new range of applications of complex analysis on tube-type symmetric domains and of the theory of Jordan algebras.

The organization of the book is as follows. The first three sections are devoted to the basic definitions concerning the Fuchs calculus and give also a thorough exposition of the fundamental facts concerning the geometry of symmetric cones. The relationship with Jordan algebras is outlined, and the general theory is illustrated by many examples. The following Sections 4 through 8 prepare the technical tools for proving the main properties of the Fuchs calculus. It is here that the noneuclidean Riemannian structure of the underlying cone is crucially used. Sections 9 through 13 constitute the core of the book, containing the fundamental results of our pseudodifferential analysis. The final four sections develop the relationship to complex analysis and group representation theory.

The reader of this book is assumed to have some familiarity with the basic ideas of pseudodifferential analysis, such as integration by parts and Fourier transform techniques. On the other hand, the methods underlying our new approach, such as the (differential-) geometric structure of symmetric cones, the fundamental facts from representation theory, and the necessary Jordan algebraic background are developed in full detail. Thus, the book is essentially self-contained in this respect.

Contents

Introduction

Many problems that come from physics or engineering and from pure mathematics (mainly harmonic analysis) give rise to partial differential equations in domains whose topological boundaries may include, besides a smooth part, all kinds of edges, corners, or conical points. A considerable supply of local models for such manifolds is given by *symmetric cones* and by products of such cones by copies of $\mathbb{R}^n$: e.g., the open half-space $\mathbb{R}^{n+1}_+$ is just a half-line times $\mathbb{R}^n$, a corner in $\mathbb{R}^3$ is the product of three half-lines, etc. Our task in the present monograph is to develop a pseudodifferential analysis exactly adapted to any of the aforementioned local models Λ.

As one of the main tools in the study of elliptic boundary value problems, pseudodifferential analysis on manifolds with a (possibly singular) boundary has been treated by a variety of authors. We shall explain presently to what extent, and why, our treatment is different from most. Meanwhile, let us note that applying pseudodifferential analysis to (elliptic or nonelliptic) boundary value problems is a subject even older than pseudodifferential analysis itself: indeed it is probably from the kernels that arose from potential-theoretic methods in the study of the Dirichlet, Neumann, and oblique-derivative problems that pseudodifferential analysis got most of its early inspiration. What gives analysis on manifolds with boundary its particular flavor is that one has to combine operators on the manifold itself with operators that act on functions defined only on the boundary, together with operators, of trace type or Poisson type, that act from one of these two species of functions to the other—all this is put together in Boutet de Monvel's algebra [B6]. A relatively recent imposing work along these lines is Grubb's [G2], which also contains a comprehensive bibliography. On a manifold with smooth boundary, it is quite natural, on the other hand, to consider *totally characteristic* operators, which degenerate in some specific way on the boundary. In a very rough way (see the end of Section 13 for more details), one may describe these as being the class of pseudodifferential operators modeled on the differential operators $P(t,x,t\,(d/dt), \partial/\partial x_j)$ where, locally near the boundary, one has distinguished a set (x_j) of tangential variables from a normal variable t. A class of totally characteristic operators was introduced by Melrose in [M1] and discussed also in Hörmander [H5]. One may note, also, that the study of particular classes of elliptic operators that degenerate in some prescribed way near a boundary was initiated in earlier works (cf. Baouendi and Goulaouic [BG], Bolley and Camus [BC]).

The study of elliptic boundary value problems on manifolds with conical singularities is a complicated subject, for which we shall refer to the experts: Rempel and Schulze [RS1, RS2], Dauge [D1], Schulze [S2], and Schrohe and Schulze [SS], with due apologies to other important contributors, who will often, however, find themselves in the lists of references of the works we have just quoted. In all these works, the conical singularities make themselves felt heavily in the description of asymptotics of solutions nearby. We take it from [D1] and [SS] that this subject was initiated by Kondrat'ev [K1]. Besides pseudodifferential analysis, the Mellin type operators (introduced by Lewis and Parenti [LP]) sometimes play a role in the methods: paraphrasing from [RS1], p. 45 and 85, these are operators somewhat similar to pseudodifferential operators with the Fourier transform replaced by a Mellin transform. However, they are quite specialized in some variable $t > 0$ (the "normal" variable above, say) and in some way not as general as one would like them to be.

One should refrain from thinking that there exists, at present, such a thing as a classification of "nice" singularities suitable for the existence of a pseudodifferential analysis. One obvious thing is that singularities quite often have a stratified structure into subsets of greater and greater codimension where the boundary is getting "more and more singular." Besides the elementary example of a corner, only think of the various pieces of the boundary of the set of all positive-definite symmetric matrices in $\mathbb{R}^n$. This is why one might advocate, as done in [SS] an "iterative" approach to the construction of such kinds of analyses. Our point of view is different—it is that harmonic analysis provides us with a wealth of such manifolds with singularities, namely the symmetric cones, which might well serve as a catalogue of what the most important types of singularities may be; admittedly, some kinds of singularities, e.g. cusps, fall out of this range. On the other hand, the task of piecing together such analyses so as to account for the analysis on manifolds whose boundaries are modeled only locally on such cones has not yet been fulfilled, but should not be too difficult a job when needed. Indeed, what has already been done here is a study of the transformation of operators on symmetric cones under appropriate changes of coordinates.

Neither should one believe that the kind of pseudodifferential analysis one has in mind is characterized by the C^∞ structure of the given manifold with boundary. To give an example (some more explanations are to be found at the end of Section 13), consider the unit ball defined by $\Sigma\, x_j^2 \leq 1$. The totally characteristic calculus mentioned earlier would lead one into modeling general (pseudodifferential) operators on differential operators that lie in the algebra [on C^∞ (closed ball)] generated by a certain set of first-order operators among which only the normal differentiation $(1 - |x|^2)\, \Sigma\, x_j\, \partial/\partial x_j$ is degenerate at the boundary: this is indeed possible, choosing $\bar{\mathbb{R}}^n_+$ as a local model. On the other hand, an equally interesting but quite distinct calculus is modeled on the first-order operators $(1 - |x|^2)\, \partial/\partial x_j$, all of which are degenerate on the sphere: this calculus is a particular case, if not of the present "Fuchs calculus" on symmetric cones, but at least of the "Klein-Gordon" calculus [U7], which is related to the Fuchs calculus on a light cone.

It is time to tell the reader that we strongly urge the use of the phrase "pseudodifferential analysis" rather than the more common one of "pseudodifferential operators." Indeed, the latter implicitly carries with it two notions: first, that only operators that do behave in some appropriate manner, asymptotically speaking, are worth considering, if not living; next, that the correspondence rule $f \mapsto Op(f)$, which assigns an operator to a "symbol," should be taken essentially for granted, it being a small variation, at the most, of the "standard" or "Weyl" calculus (see Section 0 for the definition of these two terms, if needed). Now we believe, on one hand, that harmonic analysis demands that the rule $f \mapsto Op(f)$ should be adapted to the given manifold (in the present work a symmetric cone), next that "all" linear operators should be granted symbols in a one-to-one way, the tamer ones only being characterized as having tamer symbols.

Let us show on the simplest possible example, that of the open half-line $\mathbb{R}^+$, why a simple restriction to $\mathbb{R}^+$ of the standard pseudodifferential analysis on the line is not appropriate. Indeed, let $f(y,\eta)$ be a function on $\mathbb{R}^+ \times \mathbb{R}$, and assume one has chosen to associate with f the operator A defined as

$$(Au)(s) = \int f(s,\eta)\, e^{2i\pi(s-t)\eta}\, u(t)\, dt\, d\eta.$$

The kernel of A is $(\mathcal{F}_2{}^{-1}f)(s,s - t)$ if $\mathcal{F}_2{}^{-1}f$ denotes the inverse Fourier transform of f with respect to the second variable. Thus, only the values of $(\mathcal{F}_2{}^{-1}f)(s,y)$ for which $y < s$ are relevant in the integral if u is defined only on $\mathbb{R}^+$. It is therefore not a surprise that this calculus shows a less than perfect behavior, in that properties of f and of A do not correspond in a simple way. The remedy consists in replacing $f(s,\eta)$, in the defining formula, by $f((st)^{1/2},\eta)$ since the map $(s,t) \mapsto ((st)^{1/2}, s - t)$ is a global diffeomorphism from $\mathbb{R}^+ \times \mathbb{R}^+$ to $\mathbb{R}^+ \times \mathbb{R}$. The resulting calculus, introduced in [U5], was called the *Fuchs calculus* of operators since differential operators that deserve, in it, to be called elliptic at zero, are just the ordinary differential operators of Fuchs type.

It may be noted at this point that it is in the same way, exactly, that one goes from the standard calculus of pseudodifferential operators to the Weyl calculus, replacing $f(s,\eta)$ by $f((s + t/2), \eta)$. Of course, the Weyl calculus is even less appropriate, as a calculus on the half-line, than the standard one. Also, one may point out now the fact that the Weyl calculus has other generalizations too, some of which shall be referred to later.

Our starting point was to realize [U8] that the definition of the Fuchs calculus can be generalized to the following cases: assume that Λ is an open subset of $\mathbb{R}^n$ endowed with an intrinsic Riemannian geometry for which it is a symmetric space, so that one can define the geodesic middle mid(s,t) of any two points s, t of Λ; assume, moreover, that the map $(s,t) \mapsto (\text{mid}(s,t), s - t)$ of $\Lambda \times \Lambda$ to $\Lambda \times \mathbb{R}^n$ is a global diffeomorphism. Now, the best-known open subsets of $\mathbb{R}^n$ that satisfy this property are just the *symmetric cones*, and it is the obvious generalization of the defining formula for (Au)(s), with s replaced by mid(s,t), that we shall call the Fuchs symbolic calculus on Λ and study in detail. In the case when Λ is the solid light cone, this program was achieved in [U6].

Symmetric cones in $\mathbb{R}^n$ carry two very important elements of structure: the Jordan-algebraic structure and the fact that the complexified space $\Pi = \Lambda + i\mathbb{R}^n$ is a (hermitian) symmetric space in its own right. The first one was overlooked in [U6] (in the light-cone case) and turns out to be useful in establishing various inequalities of a geometric nature. On a deeper level, however, one might argue that Jordan algebras have been known to have a role to play in quantum mechanics for a long time: indeed, they account, essentially, for the symmetric part (1/2) (AB + BA) in the product of two operators, whereas the much more familiar Lie algebras account for the antisymmetric part. In this direction, infinite-dimensional Jordan C*-algebras do play a role too, as has been shown in [U12] and [U13].

The role of the complexified space Π deserves to be explained in greater detail. A symmetric cone Λ in $\mathbb{R}^n$ is acted upon in a transitive way under the group $GL(\Lambda)$ of all linear transformations of $\mathbb{R}^n$ that preserve it. Choosing on Λ a $GL(\Lambda)$-invariant measure, one obtains a unitary representation of $GL(\Lambda)$ in $L^2(\Lambda)$; it extends as a unitary representation U of G, a semi-direct product of $GL(\Lambda)$ by $\mathbb{R}^n$, if one lets $\tau \in \mathbb{R}^n$ act on $L^2(\Lambda)$ as the multiplication by $e^{2i\pi\langle\tau,t\rangle}$. By its very construction, the Fuchs calculus is covariant under this representation and under the action of G on $\Lambda \times \mathbb{R}^n$ given by

$$[P,\tau] \cdot (y,\eta) = (Py, P'^{-1}\eta + \tau),$$

in which P′ is the transpose of $P \in GL(\Lambda)$. This means that the operator with symbol $f \circ [P,\tau]^{-1}$ is just $U(P,\tau)\, A\, U(P,\tau)^{-1}$ if A is the operator with symbol f. It is not surprising that this covariance property extends to the group Γ generated by G and T^*S, where G is identified with a group of transformations of $\Lambda \times \mathbb{R}^n = T^*(\Lambda)$, and T^*S is the cotangent map associated with the geodesic symmetry S around some point e of Λ. Note that the dimension of Γ is n plus that of G—the extension of U to a unitary representation U_∞ of Γ is the one associated with the natural action of S on $L^2(\Lambda)$.

The complex structure of $\Pi = \Lambda + i\mathbb{R}^n$, whose real underlying structure is best thought of as that of $T(\Lambda)$, will make it possible to enrich the picture in a useful way. Let G act on Π under the rule

$$(P,b) \cdot z = P'^{-1} z - ib.$$

Let the invariant measure dm on Λ be connected to the Lebesgue measure dt by $dm(t) = \omega(t)^{-1}\, dt$. Then the measure

$$d\mu(z) = \omega(\text{Re } z)^{-2}\, d(\text{Re } z)\, d(\text{Im } z)$$

is invariant on Π. On the other hand, the Jordan-theoretic structure associated with Λ makes it natural to define a certain polynomial Δ on $\mathbb{R}^n$, which is positive on Λ. In the case when Λ is an irreducible symmetric cone, ω is a power of Δ. For any real number λ, and $z \in \Pi$, define the function ψ_z^Λ on Λ by

$$\psi_z^\lambda(t) = (\Delta(\text{Re } z)\, \Delta(t))^{\lambda/2}\, e^{-2\pi<t,z>}$$

so that the functions ψ_z^λ are permuted under the representation U of G:

$$U(\gamma)\psi_z^\lambda = \psi_{\gamma\cdot z}^\lambda \quad (\gamma \in G,\ z \in \Pi).$$

Let $(\,|\,)$ denote the scalar product on the Hilbert space $L^2(\Lambda)$, complex-linear in the second variable: then, if $\lambda > \lambda_0$ (depending on Λ), there exists some constant $c_\lambda > 0$ with

$$\int_\Pi (u|\psi_z^\lambda)\, (\psi_z^\lambda|v)\, d\mu(z) = c_\lambda (u|v)$$

for all $u,v \in L^2(\Lambda)$. In view of the last two formulas, we call (ψ_z^λ), for fixed λ, a set of *coherent states.* The terminology is borrowed from the physicists, who usually have a somewhat narrower concept (see, however, Perelomov [P1]). As a tool in pseudodifferential analysis, coherent states were introduced by one of the authors in [U3] and all subsequent papers.

Indeed, they make it possible to reduce the study of an operator A to that of the function $(\psi_z^\lambda | A\psi_w^\lambda)$ of z,w. As a consequence, ugly operations like that of cutting symbols into pieces can usually be dispensed with, especially (but not only), when a group is present; also, the procedures usual in the pseudodifferential analysis on $\mathbb{R}^n$ could not work here as they depend on an integral composition formula that is much too complicated in our case.

The main task in this work consists in giving characterizations of classes of symbols in terms of estimates for $(\psi_z^\lambda | A\psi_w^\lambda)$, A being the associated operator. For instance, group invariance properties make the definition of a class of symbols of uniform type and weight 1 quite natural—it is just the analogue of what is, in the pseudodifferential analysis on $\mathbb{R}^n$, the Calderon-Vaillancourt class $S_{0,0}^0$. Let d be the geodesic distance on Π associated with its hermitian structure; then, the operators A with symbols of the type above are just those that satisfy the *main estimate*

$$|(\psi_z^\lambda | \, A\psi_w^\lambda)| \leq C_N \, e^{-Nd(z,w)} \tag{ME}$$

for all N, and λ large enough. Once this has been proved (it is quite a task), many results follow without difficulty. The continuity properties of operators, the fact that they constitute algebras, the asymptotic expansion of product symbols (see [U6; Section 1] or the present Section 0 for a short presentation of the Weyl calculus on $\mathbb{R}^n$ along those lines). Also, we shall rely on coherent states (of a somewhat more general nature) to give a characterization *à la* Beals (cf. [B2]) of a certain class of operators A. This means a characterization in terms of the iterated brackets of A with the infinitesimal operators of the representation U. This characterization, as in the pseudodifferential operator theory on $\mathbb{R}^n$, has nice consequences, among them the change of coordinates analysis needed for an extension of the calculus to manifolds.

It may be useful, at this point, to explain why, in order to obtain the estimates for $(\psi_z^\lambda | A\psi_w^\lambda)$, we have not followed the method used in [U6] for the light-cone case, as it might appear as the most natural one. Let us define the concept of Wigner function of a pair of functions u, v as the function W(u,v) on $T^*(\Lambda)$ such that

$$(u | Av) = \int f(y,\eta) \, W(u,v)(y,\eta) \, dy \, d\eta$$

if A is the operator with symbol f. Then, obviously, what we have to do is just to give estimates for W(u,v); by the way, the concept of Wigner functions makes sense for any symbolic calculus. As it turns out, $W(\psi_z^\lambda, \psi_w^\lambda)$ is a pleasant special function: however, our experience with the light-cone case has shown that estimates for this function depend on quite tricky geometric inequalities on Π, whose proof may be even harder in the general case. This is why we have switched to another kind of proof, involving the use of a one-parameter family of symbols, the μ-symbols of A. This splits the difficulty of proving (ME) in two, namely one has to show that it is true if A has a nice μ-symbol for all μ, besides proving that the μ symbols of A are nice if the Fuchs symbol is. As it turns out, this makes the proof of (ME) much simpler and less Λ-dependent: the μ-symbol trick was introduced in the Klein-Gordon calculus of operators [U7], a symbolic calculus on $\mathbb{R}^n$ whose relationship to relativistic mechanics is the same as that of the Weyl calculus to nonrelativistic mechanics, and a descendant of the Fuchs calculus on the light cone; the parameter μ, whose role is purely technical, has nothing to do with the parameter λ to be introduced now.

The space Π, being a hermitian symmetric space, has a group Aut(Π) of complex automorphisms (also isometries) much larger than G, since one can add to G the symmetry around any point of Π: observe that Aut(Π) is not the same as Γ, in which geodesic symmetries on Λ, rather than Π, were used. A known Paley-Wiener theorem [K2, FK2] shows that, for λ large enough, a certain (weighted) Laplace transform is an isometry from $L^2(\Lambda)$ onto the space $H^2_\lambda(\Pi)$ of holomorphic functions f on Π satisfying

$$\int_\Pi |f(z)|^2 \,\Delta(\mathrm{Re}\, z)^\lambda \, d\mu(z) < \infty.$$

Now $H^2_\lambda(\Pi)$ is just the Hilbert space of a projective unitary representation U_λ of Aut(Π): namely (cf., e.g., Rossi and Vergne [RV] or Gross and Kunze [GK]) the family (U_λ) is the so-called holomorphic discrete series of representations of Aut(Π) (no longer discrete when projective representations are allowed). The restriction of U_λ to G is (up to equivalence) independent of λ and coincides with the representation U discussed before. As λ goes to infinity, U_λ contracts to the representation U_∞ of Γ, the full Fuchs covariance group, in the following sense—the infinitesimal operators of U_λ, as λ goes to infinity, converge (after some renormalization) to those of Γ. Since Π is complex, Aut(Π) has a natural circle subgroup S^1, and the infinitesimal generator L_λ of the corresponding unitary group is the analogue, in the H_λ-theory, of the harmonic oscillator well-known on $\mathbb{R}^n$. Then, to show that U_λ contracts to U_∞, one just has to compute the limit of L_λ.

Even though this goes beyond the range of the present work, let us mention that there exists another generalization of the Weyl calculus, based on the use of (H^2_λ (Π)), U_λ) and on that of Π as a phase space, covariant under the actions of Aut(Π) on $H^2_\lambda(\Pi)$ and Π. It is the one that makes use, as building blocks, of the unitary self-adjoint operators associated under U_λ to the symmetries on Π. This calculus, formally introduced in [U4] in the general frame of a quantization program for hermitian symmetric spaces endowed with suitable line bundles, was studied in detail in the case when Aut(Π) = SL(2,$\mathbb{R}$) in [U9] and [U11], in which rather complete results were obtained; much less complete results were obtained in [U10] in the SO(2,n + 1) case, and in [U16] for the SU(r,r) case. It is part of the authors' intention to carry this study further, in the general case, since a great deal of interesting harmonic analysis can be obtained from it. As a pseudodifferential analysis, however, this "λ-Weyl" calculus is never as good as the Fuchs calculus, even though it becomes as decent as one pleases as λ increases. As a matter of fact, the Fuchs calculus, in the half-line case, was discovered precisely in this way, as a calculus meant to smooth out the difficulties arising in the λ-Weyl calculus from the finiteness of λ; then, it is not surprising that the correct representation to use should be the limit of U_λ as λ goes to infinity! However, one should refrain, in the Fuchs calculus, from any overemphasis on this scheme: for instance, it is at some points necessary to use coherent states more general than the functions ψ_z^λ naturally associated with $H^2_\lambda(\Pi)$.

Even though the present work does not contain applications to partial differential equations, one of the *raisons d'être* of the Fuchs calculus (cf. [U5] for the half-line case) lies of course in this realm of investigations. From its very construction, the Fuchs calculus is connected to harmonic analysis, and to the operators that originate from it, in a more thorough way than any other pseudodifferential analysis on a symmetric space. Indeed, it is the full Riemannian structure of such a space, together with its group of isometries, and not only the one-dimensional subgroup of dilations, that plays a role there. However, even on a half-space $\bar{\mathbb{R}}^n_+$, the use of what we deem is the correct rule f $\mapsto$ Op(f) allows, as should be expected, for the description of classes of operators much more general than, say, the usual totally characteristic calculus (cf. end of Section 13 for a comparison with Melrose's calculus). Also, it has been pointed out to us that the Fuchs calculus on a half-line has been used now and then in signal analysis, where its associated concept of Wigner function has turned out to give a better time-frequency description of a signal than the more traditional one (this is due to the fact that frequency is, by essence, positive). Finally, as a manageable limit of the (Weyl) U_λ-calculus referred to above, the Fuchs calculus will play a technically important role in the study of the (curved-phase space) Weyl calculus itself, which it is part of the authors' intentions to carry along the lines of what was done in [UU] for the Berezin calculus.

Acknowledgments

The authors would like to thank Todd Kirkham, Sharon Gumm, and Monika Teubner for the expert typing of the manuscript. The first named author (A.U.) was partially supported by the CNRS (URA 1870), while the second named author (H.U.) was partially supported by the National Science Foundation (DMS 900-2958).

0

Pseudodifferential Analysis on $\mathbb{R}^n$

This introductory section was added only as an afterthought and contains no novel features. Its principal aim is to introduce, in the much more familiar case of pseudodifferential operators on $\mathbb{R}^n$, some of the methods that shall be generalized later to symmetric cones. Indeed, even on $\mathbb{R}^n$, these methods, which owe much to ideas from harmonic analysis and from physics, are quite different from those used by most authors on the subject. This section should make the rest of the book more accessible to the reader. Its emphasis is entirely on the building of pseudodifferential analysis, not on how to use it. For this, the reader is urged to consult Taylor's book [T2] or the first of the two volumes by Treves [T1]; other treatments include the book [K5] by Kumano-go and the third volume [H5] in Hörmander's series, in particular Chapter 18.

One possible way of introducing pseudodifferential operators on $\mathbb{R}^n$ is, as their name should indicate, as a generalization of linear differential operators. For convenience, let us deal only, temporarily, with operators whose coefficients a_α ($|\alpha| \le m$) belong to Schwartz' space $\mathcal{S}(\mathbb{R}^n)$ of rapidly decreasing C^∞ functions on $\mathbb{R}^n$. Then, for $u \in \mathcal{S}(\mathbb{R}^n)$ and $x \in \mathbb{R}^n$,

$$\begin{aligned}(Au)(x) &= \sum a_\alpha(x)(D^\alpha u)(x)\\ &= \sum a_\alpha(x) \int \xi^\alpha e^{2i\pi\langle x-y,\xi\rangle} u(y)\, dy\, d\xi\\ &= \int a(x,\xi) e^{2i\pi\langle x-y,\xi\rangle} u(y)\, dy\, d\xi \qquad (0.1)\end{aligned}$$

(each of the two integrals above, as well as (0.3) and (0.6) below, should be thought of as a superposition—dy first) if one introduces the function

$$a(x,\xi) = \sum a_\alpha(x)\xi^\alpha, \tag{0.2}$$

called the *standard symbol* of A : note that, as a function of ξ, it is a polynomial. Also, note that applying, in the usual sense, the operator $a_\alpha D^\alpha$ means applying

$$D^\alpha = \left(\frac{1}{2i\pi}\frac{\partial}{\partial x}\right)^\alpha$$

first, then the multiplication by a_α. The reverse choice would have been equally possible and would have led to substituting for A the differential operator $\tilde{A}$ with

$$(\tilde{A}u)(x) = \int a(y,\xi)\, e^{2i\pi\langle x-y,\xi\rangle}\, u(y)\, dy\, d\xi. \tag{0.3}$$

Then, a is called the *antistandard symbol* of $\tilde{A}$. Of course, expanding $D^\alpha(a_\alpha u)$ under Leibniz rule, one sees that $\tilde{A}$ also has a standard symbol; it is not difficult, either, to check that the standard symbol of $\tilde{A}$ is given by

$$\begin{aligned} Ja &:= \exp\left(\frac{1}{2i\pi}\sum\frac{\partial^2}{\partial x_j\partial\xi_j}\right)a \\ &= \sum_{N\geq 0}\frac{1}{N!}\left(\frac{1}{2i\pi}\sum\frac{\partial^2}{\partial x_j\partial\xi_j}\right)^N a, \end{aligned} \tag{0.4}$$

a series with a finite number of nonzero terms only since, as a function of ξ, a is a polynomial. A symbolic calculus of *differential operators* is just an expansion of Leibniz rule, to wit, a set of rules, as simple as possible, for computing the symbol of the composition AB of two given differential operators or that of the formal adjoint A^*, on $L^2(\mathbb{R}^n)$, of A. For instance, if A and B have standard symbols a (= $a(x, \xi)$) and b respectively, that of AB is $a\square b$, with

$$(a\square b)(x,\xi) = \sum\frac{1}{\beta!}(2i\pi)^{-|\beta|}\partial_\xi^\beta a(x,\xi)\cdot\partial_x^\beta b(x,\xi) \tag{0.5}$$

(to check it, one may consider only the case when $a(x,\xi)$ reduces to ξ^α for some $\alpha \in \mathbb{N}^n$). The standard symbol of A^* can, as $A^*u = \Sigma\, D^\alpha\, (\bar{a}_\alpha u)$, be obtained from (0.4) since A^* appears directly as the operator whose antistandard symbol is the function $(x,\xi) \mapsto \bar{a}(x,\xi)$. It may look regrettable that we get the simplest conceivable formula only if we make the self-contradictory choice of representing operators A by their standard symbols, and their formal adjoints by their antistandard symbols. One way out of this dilemma was proposed by H. Weyl and amounts to choosing neither the standard nor the antistandard symbol but letting them meet halfway. Thus, the operator Op(f) with *Weyl symbol* f is defined through

$$(\mathrm{Op}(f)u)(x) = \int f\left(\frac{x+y}{2}, \xi\right) e^{2i\pi<x-y,\xi>} u(y)\, dy\, d\xi. \tag{0.6}$$

Then, the Weyl symbol of $\mathrm{Op}(f)^*$ is just $\bar{f}$. The Weyl symbol f and the standard symbol a of the same differential operator are related by $a = J^{1/2}f$, where one can set, generally (cf. (0.4)),

$$J^t = \exp\left(\frac{t}{2i\pi}\sum \frac{\partial^2}{\partial x_j \partial \xi_j}\right) = \sum \frac{1}{N!}\left(\frac{t}{2i\pi}\sum \frac{\partial^2}{\partial x_j \partial \xi_j}\right)^N = \sum \frac{1}{\alpha!}\left(\frac{t}{2i\pi}\right)^{|\alpha|} \partial_x^\alpha \partial_\xi^\alpha. \tag{0.7}$$

Combining (0.5) and (0.7), one gets the Weyl symbol f#g of AB, if A and B have Weyl symbols f and g, as

$$f\#g = J^{-1/2}((J^{1/2}f) \,\square\, (J^{1/2}g)) \tag{0.8}$$

or, expressing the result as the restriction to the diagonal of $\mathbb{R}^{2n} \times \mathbb{R}^{2n}$ of the image of $f \otimes g$ under the appropriate operator apparent from (0.8),

$$(f\#g)(x,\xi) = \exp \frac{1}{4i\pi}\sum\left(-\frac{\partial^2}{\partial y_j \partial \zeta_j} + \frac{\partial^2}{\partial z_j \partial \eta_j}\right)(f(y,\eta)g(z,\zeta)), \tag{0.9}$$

where the whole right-hand side must be evaluated at $(y,\eta) = (z,\zeta) = (x,\xi)$. This can also be written as

$$(f\#g)(x,\xi) = \sum \frac{(-1)^{|\alpha|}}{\alpha!\beta!}\left(\frac{1}{4i\pi}\right)^{|\alpha|+|\beta|} \partial_x^\alpha \partial_\xi^\beta f(x,\xi) \cdot \partial_x^\beta \partial_\xi^\alpha g(x,\xi). \tag{0.10}$$

Compared to (0.5), this may look as a deterioration. However, this is not the case. Indeed, if the terms on the right-hand sides with $\beta = 0$ (resp. $\alpha = \beta = 0$) reduce to ab (or fg) in both cases, those with $|\beta| = 1$ (resp. $|\alpha| + |\beta| = 1$) are respectively $(2i\pi)^{-1} \sum (\partial a/\partial \xi_j)(\partial b/\partial x_j)$ and

$$(4i\pi)^{-1}\{f,g\} = (4i\pi)^{-1}\sum\left[\frac{\partial f}{\partial \xi_j}\frac{\partial g}{\partial x_j} - \frac{\partial f}{\partial x_j}\frac{\partial g}{\partial \xi_j}\right], \tag{0.11}$$

where the Poisson bracket { , } familiar from Hamilton's mechanics appears. It is canonically associated with the symplectic form on

$\mathbb{R}^{2n} = \mathbb{R}^n \times \mathbb{R}^n$, by which is meant the antisymmetric bilinear form [,] on $\mathbb{R}^{2n} \times \mathbb{R}^{2n}$ defined by

$$[(x,\xi), (y,\eta)] = -\langle x,\eta\rangle + \langle y,\xi\rangle. \tag{0.12}$$

It is time to tell what *pseudodifferential* operators are. The Weyl calculus is defined just by (0.6), except that now symbols f that are no more polynomials as functions of ξ are allowed. One could also define an operator with given standard symbol a by a generalization of the last integral on the right-hand side of (0.1). The link between the Weyl symbol f and the standard symbol a of the same operator is still given by $a = J^{1/2}f$ with J^t as in (0.7), only now the exponential of the operator $(2i\pi)^{-1} \Sigma\, \partial^2/\partial x_j\, \partial\xi_j$ should not be defined by means of a series but with the help of the Fourier transformation. We have not yet told what kind of assumptions one should make about f to make (0.6) meaningful. Actually Op(f) makes sense, as a weakly continuous operator from $\mathcal{S}(\mathbb{R}^n)$ to its dual space $\mathcal{S}'(\mathbb{R}^n)$, whenever $f \in \mathcal{S}'(\mathbb{R}^n \times \mathbb{R}^n)$, i.e., f is a tempered distribution on $\mathbb{R}^{2n}$; also, when $f \in \mathcal{S}(\mathbb{R}^{2n})$, Op(f) extends as an operator on $\mathcal{S}'(\mathbb{R}^n)$, and maps the whole of it to $\mathcal{S}(\mathbb{R}^n)$. We shall not prove these two (trivial) facts, concerned with soft spaces like $\mathcal{S}(\mathbb{R}^n)$ and $\mathcal{S}'(\mathbb{R}^n)$. Before we come back to some harder pseudodifferential analysis on $\mathbb{R}^n$, we digress to tell briefly why physicists have been interested in the Weyl symbolic calculus.

In the late 1920s, foundational axiomatics for quantum mechanics had been set by von Neumann. One of the two major processes involved was that of measurement—the general idea was that, starting from some *phase space*, the manifold suitable for the description of some classical system, one should be able to define, as canonically as possible, a Hilbert space H and a correspondence rule $f \mapsto \mathrm{Op}(f)$ from functions on the phase space (the *classical observables*) to operators on H, generally unbounded. In the case we are interested in, the phase space would be $\mathbb{R}^n \times \mathbb{R}^n$ and H would be $L^2(\mathbb{R}^n)$ (then, $\mathbb{R}^n$, on which functions in H live, is called the *configuration space*, and the second copy of $\mathbb{R}^n$, on which the Fourier transform of $u \in L^2(\mathbb{R}^n)$, defined as

$$\hat{u}(\xi) = \int u(x)\, e^{-2i\pi<x,\xi>}\, dx, \tag{0.13}$$

lives, is called the *momentum space*). The Weyl rule Op, as defined in (0.6), was proposed by H. Weyl [W3] in 1928, precisely so as to fulfill the preceding quantization program.

There is a concept dual to Weyl's rule $f \mapsto \mathrm{Op}(f)$, introduced by E. Wigner four years later (for other reasons, as an approach to the nonexisting probability measure on phase-space that would make quantum mechanics a chapter of probability theory; the "Wigner function" was rediscovered time and again by various people; one should mention the entertaining paper [D2] by De Bruijn). The Wigner function $W(\psi,\varphi)$ of any two functions $\varphi, \psi \in \mathcal{S}(\mathbb{R}^n)$ (physicists would consider only the case when $\psi = \varphi$) is that function on $\mathbb{R}^n \times \mathbb{R}^n$, which makes the identity

$$(\psi \mid \mathrm{Op}(f)\varphi) = \int f(y,\eta)\, W(\psi,\varphi)(y,\eta)\, dy\, d\eta \tag{0.14}$$

valid for every $f \in \mathcal{S}'(\mathbb{R}^n)$. Here the left-hand side, defined as an extension of the scalar product of $L^2(\mathbb{R}^n)$ [but here $\mathrm{Op}(f)\varphi$ lies in $\mathcal{S}'(\mathbb{R}^n)$] is linear with respect to φ, antilinear with respect to ψ. It is clear from (0.14) that $W(\psi,\varphi) \in \mathcal{S}(\mathbb{R}^n \times \mathbb{R}^n)$, and, starting from (0.6), one obtains the formula

$$W(\psi,\varphi)(y,\eta) = 2^n \int \varphi(y + x)\, \bar{\psi}(y - x)\, e^{-4i\pi<x,\eta>}\, dx. \tag{0.15}$$

One may also note, as the consequence of another purely formal computation, that $W(\psi,\varphi)$ happens to be, too, the symbol of the rank-one operator $u \mapsto (\psi,u)\varphi$.

Another concept that owes its current terminology to physicists (but its existence to Laplace) is that of a *family of coherent states*: one may look at Perelomov's book [P1] for some fairly general account. There is no clear-cut universal definition, but here is the rough idea. Given a Hilbert space H, a family of coherent states in H is a family (ψ_τ) of elements of H parametrized by points τ in some measure space, satisfying the following: there exists a constant $C > 0$ such that

$$C^{-1}\, \|u\|^2 \leq \int |(\psi_\tau \mid u)|^2 d\tau \leq C\, \|u\|^2 \tag{0.16}$$

for every $u \in H$. In the (frequent) case when H can be realized either as a space of functions on some configuration space or as a space of functions on the dual momentum space, the two being connected by some Fourier transformation, one demands, moreover, that in both realizations ψ_τ should decay at infinity: thus, when $H = L^2(\mathbb{R}^n)$, the family of plane waves $x \mapsto \exp(-2i\pi\, \langle x,\tau\rangle)$, $\tau \in \mathbb{R}^n$, would *not* qualify even though (0.16) works (actually, there is also, in this case, the fact that a plane wave does not lie in $L^2(\mathbb{R}^n)$). Usually the family (ψ_τ) is *overcomplete,* i.e., quite far from being linearly independent. *Wavelet* theory (we shall have no use for it) concerns itself with cases where orthogonality is recovered (then, the parametrizing space is discrete). It is quite easy to build families of coherent states in special situations from the Laplace transform, or from Parseval's formula so as to integrate out half of the variables in (0.16) (this is the most common recipe), or from the theory of square-integrable representations (cf. [W1; p. 351]). Much more difficult constructions (on which we shall report later on in this section) use pseudodifferential analysis, but, at the present time, the game will be played the other way around, coherent states acting as a tool in pseudodifferential analysis. Start from any function $\varphi \in \mathcal{S}(\mathbb{R}^n)$, with $\|\varphi\| = 1$ in the L^2-sense. As a parametrizing space, take $\mathbb{C}^n = \{z = x + i\xi\}$ with its standard Lebesgue measure $d\mu(z)$. For every $z \in \mathbb{C}^n$, define $\varphi_z \in \mathcal{S}(\mathbb{R}^n)$ by

$$\varphi_z(t) = \varphi(t - x) \exp 2i\pi\langle t - \frac{x}{2}, \xi\rangle. \tag{0.17}$$

Observe that

$$(\varphi_z)_{z'}(t) = \varphi_{z+z'}(t)\, e^{-i\pi \operatorname{Im}\langle z,\bar{z}'\rangle}, \tag{0.18}$$

where $\operatorname{Im}\langle z,\bar{z}'\rangle$ may also be written as $[(x,\xi), (x',\xi')]$ by reference to the symplectic form given by (0.12). Since $\exp(-i\pi \operatorname{Im}\langle z,\bar{z}'\rangle)$ is a constant (i.e., does not depend on t) of modulus 1, (0.18) means that, under (0.17), the additive group $\mathbb{C}^n$ acts on $L^2(\mathbb{R}^n)$ as a *projective* representation, i.e., as a representation up to phase factors: this is of course the Heisenberg representation. It is immediate (e.g., from Parseval's formula) that (0.16) is valid with $C = 1$. By polarization, one has the identity

$$(v\,|\,u) = \int_{\mathbb{C}^n} (v\,|\,\varphi_z)(\varphi_z\,|\,u)\, d\mu(z) \tag{0.19}$$

for every pair of functions in $L^2(\mathbb{R}^n)$. It is easy to see, using the identities

$$z_j\varphi_z(t) = \left(t_j + \frac{1}{2\pi}\frac{\partial}{\partial t_j}\right)(\varphi_z(t)) - \left(t_j\varphi + \frac{1}{2\pi}\frac{\partial\varphi}{\partial t_j}\right)_z(t),$$

$$\bar{z}_j\varphi_z(t) = \left(t_j - \frac{1}{2\pi}\frac{\partial}{\partial t_j}\right)(\varphi_z(t)) - \left(t_j\varphi - \frac{1}{2\pi}\frac{\partial\varphi}{\partial t_j}\right)_z(t) \tag{0.20}$$

and setting, for $k \in \mathbb{Z}$, and $u \in \mathcal{S}'(\mathbb{R}^n)$,

$$I_k(u) = \int (1 + |z|^2)^k\, |(\varphi_z|u)|^2\, d\mu(z) \tag{0.21}$$

(where we have extended the scalar product $(\,|\,)$ as a separately continuous sesquilinear form on $\mathcal{S}(\mathbb{R}^n) \times \mathcal{S}'(\mathbb{R}^n)$) that, given any $u \in \mathcal{S}'(\mathbb{R}^n)$, one has $I_k(u) < \infty$ for some $k \in \mathbb{Z}$; also, $I_0(u) < \infty$ if and only if $u \in L^2(\mathbb{R}^n)$, and $I_k(u) < \infty$ for all k if and only if $u \in \mathcal{S}(\mathbb{R}^n)$.

The Heisenberg representation is related to the Wigner function concept as characterized in (0.14) or (0.15) as follows. With $\varphi, \psi \in \mathcal{S}(\mathbb{R}^n)$, set $\Phi(y + i\eta) = W(\psi,\varphi)(y,\eta)$: then

$$W(\psi_{z'},\varphi_z)(y,\eta) = \Phi\left(y + i\eta - \frac{z + z'}{2}\right)$$

$$\exp(2i\pi \operatorname{Im}\langle y + i\eta, \bar{z}' - \bar{z}\rangle) \exp - i\pi \operatorname{Im}\langle z,\bar{z}'\rangle \qquad (0.22)$$

for every pair (z,z') of points in $\mathbb{C}^n$.

An especially nice family of coherent states is obtained if one starts from $\varphi(t) = 2^{n/4} \exp(-\pi|t|^2)$, the normalized ground state (i.e., eigenfunction with lowest eigenvalue $n/2$) of the *harmonic oscillator*

$$L = \pi \sum (t_j^2 + D_j^2) = \operatorname{Op}(\pi(|y|^2 + |\eta|^2)) \qquad (0.23)$$

(note that we have shifted to (y,η) to denote points on the phase space $\mathbb{R}^n \times \mathbb{R}^n$, and to t on the configuration space; this is to save (x,ξ), with $x + i\xi = z$, as a parameter for the family of coherent states). Then φ_z is a ground state of

$$L_z = \operatorname{Op}((y,\eta) \mapsto \pi(|y - x|^2 + |\eta - \xi|^2)), \qquad (0.24)$$

the *harmonic oscillator centered at* z. What is nice, among other things, is that, in that case, $W(\varphi,\varphi)$ is also a Gaussian function, namely

$$W(\varphi,\varphi)(y,\eta) = 2^n \exp(-2\pi(|y|^2 + |\eta|^2)). \qquad (0.25)$$

The realization given by $u \mapsto \tilde{u}$, with $\tilde{u}(z) = (\varphi_z|u)$, of $L^2(\mathbb{R}^n)$ as a subspace of $L^2(\mathbb{C}^n)$ (namely that which consists of those functions that become antiholomorphic after having been multiplied by $\exp((\pi/2)|z|^2)$ is called the *Bargmann-Fock realization* (cf. e.g., [I1]).

We now come back to our true subject, pseudodifferential analysis on $\mathbb{R}^n$, and can start with the technicalities.

DEFINITION 0.26 A weight-function is any function $m > 0$ on $\mathbb{R}^n \times \mathbb{R}^n$, satisfying for some pair (C_1, N_1) with $C_1 > 0$ and $N_1 \geq 0$ the inequality

$$m(x,\xi) \leq C_1\, m(y,\eta)\, [1 + |x - y|^2 + |\xi - \eta|^2]^{N_1}$$

for every pair $((x,\xi), (y,\eta))$ of points of $\mathbb{R}^n \times \mathbb{R}^n$. A function $f \in C^\infty(\mathbb{R}^n \times \mathbb{R}^n)$ shall then be called a symbol of weight m if the function

$$m(y,\eta)^{-1} (\partial/\partial y)^\alpha (\partial/\partial\eta)^\beta f(y,\eta)$$

is bounded on $\mathbb{R}^n \times \mathbb{R}^n$ for every pair (α,β) of multi-indices $\in \mathbb{N}^n$.

THEOREM 0.27
Let f *be a symbol of weight* m, *and let* $\varphi,\psi \in \mathcal{S}(\mathbb{R}^n)$: *then, for each number* N, *there exists* $C > 0$ *such that the estimate*

$$|(\psi_{z'}|\mathrm{Op}(f)\varphi_z)| \le C(1 + |z - z'|^2)^{-N}\, \tilde{m}\left(\frac{z + z'}{2}\right)$$

holds for every pair (z,z′) *of points of* $\mathbb{C}^n$, *with* $\tilde{m}(x + i\xi) = m(x,\xi)$. *Conversely, let* $f \in \mathcal{S}'(\mathbb{R}^n \times \mathbb{R}^n)$ *and let* $\varphi \in \mathcal{S}(\mathbb{R}^n)$ *be arbitrary but not identically zero: set* $\psi = \varphi$ *too and assume that the above family of estimates holds; then* f *is a symbol of weight* m.

PROOF From (0.14) and (0.22) one gets

$$|(\psi_{z'}|\mathrm{Op}(f)\varphi_z)| = \left|\int f(y,\eta)\Phi\left(y + i\eta - \frac{z + z'}{2}\right)\right. \\ \left.\exp\,(2i\pi\,\mathrm{Im}\langle y + i\eta, \bar{z}' - \bar{z}\rangle)\, dy\, d\eta\right|. \qquad (0.28)$$

A standard integration by parts based on

$$\left[1 - (4\pi^2)^{-1}\sum\left(\frac{\partial^2}{\partial y_j^2} + \frac{\partial^2}{\partial \eta_j^2}\right)\right] e^{2i\pi\,\mathrm{Im}\langle y+i\eta,\bar{z}'-\bar{z}\rangle} \\ = [1 + |z' - z|^2]\, e^{2i\pi\,\mathrm{Im}\langle y+i\eta,\bar{z}'-\bar{z}\rangle}, \qquad (0.29)$$

together with the inequality

$$m(y,\eta) \le C_1\tilde{m}\left(\frac{z+z'}{2}\right)\left[1 + \left|y + i\eta - \frac{z + z'}{2}\right|^2\right]^{N_1} \qquad (0.30)$$

yields the first half of Theorem 0.27. In the reverse direction, we may assume $\|\varphi\| = 1$ so that, as a consequence of (0.19), used twice, one has

$$(v|\mathrm{Op}(f)u) = \int (v|\varphi_{z'})(\varphi_{z'}|\mathrm{Op}(f)\varphi_z)(\varphi_z|u)\, d\mu(z)\, d\mu(z') \qquad (0.31)$$

for all $u,v \in \mathcal{S}(\mathbb{R}^n)$, an identity that expresses Op(f) as an integral superposition of the rank-one operators $u \mapsto (\varphi_z|u)\varphi_{z'}$: since the symbol of such an operator is $W(\varphi_z,\varphi_{z'})$, one may write

$$f(y,\eta) = \int (\varphi_{z'}|\mathrm{Op}(f)\varphi_z)\, W(\varphi_z,\varphi_{z'})(y,\eta)\, d\mu(z)\, d\mu(z'). \qquad (0.32)$$

This time, it is sufficient to differentiate under the integral sign, using the expression (0.22) of the Wigner function involved and the assumption about $(\varphi_{z'}|\mathrm{Op}(f)\varphi_z)$ to take care of the extra powers of $z - z'$ or $\bar{z} - \bar{z}'$

that show up when differentiating: also note that (0.30) must be used in the other direction; it is reversible, which concludes the proof of Theorem 0.27. Q.E.D.

COROLLARY 0.33
For every symbol f *of weight* m, Op(f) *acts continuously from* $\mathcal{S}(\mathbb{R}^n)$ *to* $\mathcal{S}(\mathbb{R}^n)$ *and extends as a continuous operator from* $\mathcal{S}'(\mathbb{R}^n)$ *to* $\mathcal{S}'(\mathbb{R}^n)$. *In the case when* $m = 1$, Op(f) *extends as a bounded operator from* $L^2(\mathbb{R}^n)$ *to* $L^2(\mathbb{R}^n)$.

PROOF Since $\tilde{m}(z)$ is bounded by $C_1(1 + |z|^2)^{N_1}$, the corollary is an immediate consequence of Theorem 0.27 together with the characterization of the spaces $\mathcal{S}(\mathbb{R}^n)$ and $L^2(\mathbb{R}^n)$ in terms of the integrals $I_k(u)$ defined in (0.21). Q.E.D

REMARK
The last part of the corollary is a particular case of the Calderon-Vaillancourt Theorem [CV].

COROLLARY 0.34
Let f_1 *(resp.* f_2*) be a symbol of weight* m_1 (resp. m_2). *Then* $f_1 \# f_2$ *is a symbol of weight* $m_1 m_2$.

PROOF With $a_j = \mathrm{Op}(f_j)$, it suffices to write

$$\begin{aligned}(\varphi_{z'} | A_1 A_2 \varphi_z) &= (A_1^* \varphi_{z'} | A_2 \varphi_z)\\ &= \int (\varphi_{z'} | A_1 \varphi_{z''})(\varphi_{z''} | A_2 \varphi_z)\, d\mu(z'') \qquad (0.35)\end{aligned}$$

and to use the characterization provided by Theorem 0.27 as well as, again, the basic property of weight functions.

COROLLARY 0.36
(Beals' characterization of operators of weight 1). *Let* A *be a weakly continuous linear operator from* $\mathcal{S}(\mathbb{R}^n)$ *to* $\mathcal{S}'(\mathbb{R}^n)$. *The following two conditions are equivalent:*

(i) one has $A = \mathrm{Op}(f)$ *for some symbol* f *of weight* 1;

(ii) denote as t_j *(resp.* D_j*) the operator that sends* u *to the function* $t \mapsto t_j\, u(t)$ *(resp.* $t \mapsto (2i\pi)^{-1}\, \partial u/\partial t_j$*), and let* $T_1, \ldots, T_k$ *be any number of operators of the preceding two kinds. Then, setting* $(\mathrm{ad}\, T)B = TB - BT$ *for any two operators* T *and* B, *the operator*

$$(\mathrm{ad}\, T_1 \cdots \mathrm{ad}\, T_k)A$$

(reducing to A *when* $k = 0$*) extends as a bounded operator from* $L^2(\mathbb{R}^n)$ *to* $L^2(\mathbb{R}^n)$.

PROOF As an immediate consequence of (0.6), it is clear that the symbols of the operators $(\mathrm{ad}\, D_j)A$ and $(\mathrm{ad}\, t_j)A$ are $(2i\pi)^{-1}\, \partial f/\partial y_j$ and $-(2i\pi)^{-1}\, \partial f/\partial \eta_j$, respectively, with $f = f(y,\eta)$, so that (i) implies (ii). Using Theorem 0.27 again, what has to be done is showing that, under the assumption (ii), the function $(z - z')^\alpha(\varphi_{z'}|A\varphi_z)$ is bounded on $\mathbb{C}^n \times \mathbb{C}^n$ for every $\alpha \in \mathbb{N}^n$. Set

$$C_j = t_j - i\, D_j, \qquad C_j^* = t_j + i\, D_j. \tag{0.37}$$

These are the so-called creation and annihilation operators that occur in (0.20). As a consequence of this latter pair of identities, one has

$$\begin{aligned}(z_j - z_j')(\varphi_{z'}\,|\,A\varphi_z) &= -(\varphi_{z'}\,|\,(\mathrm{ad}\, C_j)A\varphi_z)\\ &\quad -(\varphi_{z'}\,|\,A(C_j\varphi)_z) + ((C_j^*\varphi)_{z'}\,|\,A\varphi_z).\end{aligned} \tag{0.38}$$

To show that (ii) implies (i), it suffices to iterate (0.38) and to estimate the right-hand side. Q.E.D.

We now pause for a few remarks, concerning all that precedes. First, everything hinges on just one theorem (Theorem 0.27), a characterization of operators of given weight by their action on coherent states. All the rest can be considered as a corollary. Also, one may note that the proof of Corollary 0.34 was particularly easy and did not depend on any integral formula for the composition of two symbols, for which we have no use. In our study of pseudodifferential analysis on symmetric cones, we shall proceed exactly along the same lines. The only difference is that the main characterization (the analogue of Theorem 0.27) will be quite difficult to prove. Another point needs to be stressed, namely that the main estimate and its converse do not depend on the weight m as much as one might figure, only on the pair (C_1, N_1) that enters Definition 0.26. Indeed, call $|||f|||_{m,N}$ the least constant C that makes

$$|\partial_y^\alpha\, \partial_\eta^\beta\, f(y,\eta)| \leq C\, m(y,\eta) \tag{0.39}$$

valid for all $(y,\eta) \in \mathbb{R}^n \times \mathbb{R}^n$ whenever $|\alpha| + |\beta| \leq N$ and, for any operator A, call $|||A|||_{m,\varphi,N}$ the least constant that makes

$$|(\varphi_{z'}|A\varphi_z)| \leq C\,\tilde{m}\,((z+z')/2)\,(1+|z-z'|^2)^{-N} \tag{0.40}$$

valid for all z, $z' \in \mathbb{C}^n$. Then, as shown by the proof of Theorem 0.27, one can, for every integer $N \geq 0$, find $\tilde{N}$ depending only on (N, N_1) and C depending only on (C_1, φ, N, N_1) such that the inequalities

$$|||\,\mathrm{Op}(f)\,|||_{m,\varphi,N} \leq C\,|||\,f\,|||_{m,\tilde{N}}, \tag{0.41}$$

$$|||\,f\,|||_{m,N} \leq C\,|||\,\mathrm{Op}(f)\,|||_{m,\varphi,\tilde{N}} \tag{0.42}$$

hold for every symbol f of weight m.

It is time to generalize to pseudodifferential symbols the composition formula that was given in (0.10) for symbols of differential operators. To that effect, we shall, from now on, call *symbols of uniform type and weight* m the symbols introduced up to now. The idea is that, when $m = 1$, the corresponding class of symbols is invariant in some uniform way under the translations of $\mathbb{R}^{2n}$; another way to put it, valid for any weight m, is to say that a symbol of uniform type and weight m can be characterized as a smooth function f with $m^{-1}f$ bounded on $\mathbb{R}^{2n}$, such that this nice behavior is not destroyed by any number of differentiations. *Symbols of classical type* are characterized by the fact that, in some very specific way, it is not only harmless, but also beneficial, to take derivatives: let us specialize the weights in that case.

DEFINITION 0.43 Given $k \in \mathbb{R}$, a symbol of classical type and order k shall be any function $f \in C^\infty(\mathbb{R}^n \times \mathbb{R}^n)$ satisfying the family of estimates

$$|\partial_y^\alpha\, \partial_\eta^\beta\, f(y,\eta)| \leq C(\alpha,\beta)\,(1+|\eta|)^{k-|\beta|}.$$

Consider the right-hand side of (0.10), assuming now that f (resp. g) is a symbol of classical type and order k_1 (resp. k_2). As a series, it need not converge. For any integer $\ell \geq 1$ one can always, however, consider the partial sum

$$S_\ell(f,g) = \sum_{|\alpha|+|\beta|<\ell} \frac{(-1)^\alpha}{\alpha!\beta!}\left(\frac{1}{4i\pi}\right)^{|\alpha|+|\beta|} \partial_y^\alpha\partial_\eta^\beta f \cdot \partial_y^\beta\partial_\eta^\alpha g \tag{0.44}$$

with $f = f(y,\eta)$, $g = g(y,\eta)$, and remark that each of the individual terms that lie further in the formal series is a symbol of classical type and order $k_1 + k_2 - \ell$. The following fact is the true justification of pseudodifferential analysis, in so much as it makes a "calculus" of pseudodifferential operators possible. Here, operators of very low order are regarded as more and more "negligible."

PROPOSITION 0.45

Under the assumptions above, $f\#g - S_\ell(f,g)$ *is a symbol of classical type and order* $k_1 + k_2 - \ell$.

PROOF As one can push ℓ as far as needed, it is enough to show that $h_\ell = f\#g - S_\ell(f,g)$ is a symbol of uniform type and weight $(1 + |\eta|)^{k_1+k_2-\ell}$. Since, as remarked in the proof of Corollary 0.36, $(2i\pi)^{-1}\ \partial f/\partial y_j$ is the symbol of $(\mathrm{ad}\ D_j)\mathrm{Op}(f)$, one may write, as a consequence,

$$\frac{\partial}{\partial y_j}(f\#g) = \frac{\partial f}{\partial y_j}\#g + f\#\frac{\partial g}{\partial y_j} \tag{0.46}$$

also, obviously

$$\frac{\partial}{\partial y_j}S_\ell(f,g) = S_\ell\left(\frac{\partial f}{\partial y_j}, g\right) + S_\ell\left(f, \frac{\partial g}{\partial y_j}\right), \tag{0.47}$$

and the same trick works with $\partial/\partial\eta_j$, so that it is sufficient to prove the estimate

$$|h_\ell(y,\eta^0)| \leq C\,(1 + |\eta^0|)^{k_1+k_2-\ell} \tag{0.48}$$

for all $(y,\eta^0) \in \mathbb{R}^n \times \mathbb{R}^n$. To this effect, a Taylor expansion of $f(y,\eta)$ with respect to η, around η^0, permits one to write

$$f(y,\eta) = (T_{\eta^\circ}^{\ell-1}f)(y,\eta) + (R_{\eta^\circ}^{\ell}f)(y,\eta), \tag{0.49}$$

where $T_{\eta^\circ}^{\ell-1}f$ is the symbol of a differential operator, and where the symbol $R_{\eta^\circ}^{\ell}f$ is a symbol of weight

$$(1 + |\eta|)^{k_1-\ell}\,(1 + |\eta - \eta^0|)^{\ell} \tag{0.50}$$

in a way that is uniform with respect to η^0. Also, note that the weight in (0.50) satisfies the basic property of weight functions set in Definition 0.26 with a pair of constants (C_1, N_1) independent of η^0. Doing the same with g in place of f, we can write $f\#g$ as the sum of four terms. Since (0.50) reduces to $(1 + |\eta^0|)^{k_1-\ell}$ at $\eta = \eta^0$, Corollary 0.34 shows that the three "remainder" terms, evaluated at η^0, are all less than $C(1 + |\eta^0|)^{k_1+k_2-\ell}$. The main term in $(f\#g)(y,\eta^0)$ arising from the decomposition is

$$(T_{\eta^\circ}^{\ell-1}f \# T_{\eta^\circ}^{\ell-1}g)(y,\eta^0)$$

and can be computed exactly from (0.10). We get all the terms in

$(S_\ell(f,g))(y,\eta^0)$, plus a bunch of extra terms that are less than $C\,(1 + |\eta^0|)^{k_1+k_2-\ell}$, which concludes the proof of Proposition 0.45.

Q.E.D.

This concludes, too, the exposition, in the case of $\mathbb{R}^n$, of the methods that are going to be used throughout the present work. We cannot, however, leave our less experienced readers with the idea that operators of classical type are the only class of pseudodifferential operators on $\mathbb{R}^n$ that one needs to be familiar with. We cannot, either, leave our more experienced readers with the wrong idea that the methods just expounded work only in the simpler cases. Pseudodifferential analysis did not start with Weyl's formula (0.6), which was largely ignored by analysts for decades. It certainly did not start as a *generalization* of differential operators, rather as a generalization of *inverses* of such. Indeed, classical potential theory on $\mathbb{R}^3$ brought into consideration operators whose integral kernels $k(x,y)$ had singularities on the diagonal comparable to that of $|x - y|^{-1}$; more generally, on $\mathbb{R}^n$, $n \geq 3$, the fundamental solution of the Laplacian is a convolution operator with a kernel like $|x - y|^{-n+2}$. To solve the Dirichlet problem in a domain of $\mathbb{R}^n$ with smooth boundary by means of a double-layer potential, one ends up with an equation for the double-layer density, on the boundary, involving an operator with some worse kind of singularity, just like $|x - y|^{-n+1}$ for a kernel on $\mathbb{R}^n$. Real troubles, and interesting things, start when one considers kernels, on $\mathbb{R}^n$, that behave near the diagonal somewhat like functions of $(x, x - y)$ homogeneous of degree $-n$ with respect to the second variable. Integrating only for $|x - y| \geq \varepsilon$, then letting ε go to zero, is a valid way of defining the associated *singular integral operator* provided one assumes the vanishing of the spherical integral $\int_{|z|=1} k_0(x,z)\, d\sigma(z)$, where $k_0(x,x - y)$ is the "principal" part of the kernel $k(x,y)$, i.e., the one that is homogeneous of degree $-n$ with respect to $x - y$. The theory of singular integral operators was tackled with by several authors, including Giraud and Mikhlin, starting from the 1930s.

Modern pseudodifferential analysis started with series of works by Calderon and Zygmund [CZ], Calderon [C1], Seeley [S3]: the paper by Calderon just quoted was foundational since, for the first time, it showed the applicability of such a kind of analysis to problems concerning *general* partial differential equations, not only the special (second-order) equations of mathematical physics. The name "pseudodifferential operators" was coined by Kohn and Nirenberg [KN], who at the same time, among others, switched from the representation of operators by singular integral kernels to the one by (standard) symbols (which Calderon and Zygmund had introduced only in an indirect way) in their very definition, involving the use of the Fourier transform. It is to be noted that the class of pseudodifferential operators considered by Kohn and Nirenberg resembled the one we discussed above as that of "operators of classical type": however, their method only allowed the consideration of symbols $f(y,\eta)$ rapidly decaying as $|y| \to \infty$ (this restriction was lifted by Calderon and Vaillancourt [CV]). Symbols of classical type and order k can be defined as those C^∞ symbols bounded by $C(1 + |\eta|^2)^{k/2}$, such that this estimate is preserved after one has applied operators $\partial/\partial y_j$ or $(1 + |\eta|^2)^{1/2}\ \partial/\partial\eta_j$ any number of times. Hörmander [H2] introduced (a y-localized version of) the so-called $S^k_{\rho,\delta}$ class, characterized as the space of symbols $f(y,\eta)$ bounded by $C(1 + |\eta|^2)^{k/2}$ to which you can apply operators like $(1 + |\eta|^2)^{\rho/2}\ \partial/\partial\eta_j$ or $(1 + |\eta|^2)^{-\delta/2}\ \partial/\partial y_j$ any number of times with no harm incurred. With this terminology, a symbol of uniform type and weight $(1 + |\eta|)^k$ (resp. of classical type and order k) would be just a symbol in the class $S^k_{0,0}$ (resp. $S^k_{1,0}$). In general, analogues of Corollaries 0.33 and 0.34 work if $\rho \geq \delta$ and $0 \leq \delta < 1$, and some form of Proposition 0.45 works if, moreover, $\rho > \delta$.

A major development occurred with the Beals and Fefferman article [BF] that put the final touch to the Nirenberg and Treves theorem relative to the local solvability of partial differential operators of principal type. Indeed, they showed that one could, in one very important instance, tailor a class of pseudodifferential operators so as to adapt it exactly to a given problem: this kind of highly versatile pseudodifferential analysis was then systematized by Beals [B1], who introduced the $S(\Phi,\varphi)$ classes: with $\Phi = \Phi(y,\eta)$, $\varphi = \varphi(y,\eta)$, the basic first-order differential operators on the phase space whose applicability, any number of times, defines the class of symbols are the operators $\Phi\ \partial/\partial y_j$ and $\varphi\ \partial/\partial\eta_j$; it is assumed that $\Phi\varphi \geq 1$, and some technical-looking, but quite essential, other assumptions are needed as well.

It may look as if pseudodifferential analysis developed, from the point of view of partial differential equations, in a way totally unrelated to physics or classical mechanics. The situation abruptly changed with Egorov's paper [E1] on the quantization of canonical transformations (which was followed up with Hörmander's theory of Fourier integral operators). A few words are in order since the underlying idea is basic to our subject, which is also connected to *quantization theory.* Consider, on the phase space $\mathbb{R}^n \times \mathbb{R}^n$, the 2-form $\Sigma\ dy_j \wedge d\eta_j$. It is, of course, closed, and C^∞ transformations of $\mathbb{R}^n \times \mathbb{R}^n$ that preserve it are called *canonical transformations.* Later on in this section, we shall have to consider the linear canonical transformations, which are also called the *symplectic transformations,* and constitute a (finite-dimensional) Lie group $Sp(n,\mathbb{R})$. From the foundations of quantum mechanics had emerged a (wrong) idea that, with some oversimplification, can be put like this: with each canonical transformation Ψ of the phase space one should be able to associate, in some more or less well-defined manner, a unitary transformation $\tilde{\Psi}$ of $L^2(\mathbb{R}^n)$ that would have the following property: given an operator A (bounded or not) on $L^2(\mathbb{R}^n)$, with symbol f (for any symbolic calculus), the operator $\tilde{\Psi}\ A\ \tilde{\Psi}^{-1}$

should have $f \circ \Psi^{-1}$ as its symbol. Now, this can be true at most in some approximate sense (i.e., $\tilde{\Psi}$ will not be unitary, only not far from invertible, and everything will work only up to error terms of low order), and what Egorov did was to show that such a construction $\Psi \mapsto \tilde{\Psi}$ is indeed possible for a large class of canonical transformations, homogeneous of degree one in the η variables.

For *some* Ψ, $\tilde{\Psi}$ can be constructed *exactly:* that this is the case whenever Ψ is a translation $(y,\eta) \mapsto (y + x, \eta + \xi)$ could of course escape no one working with pseudodifferential operators; the associated $\tilde{\Psi}$'s then generate a version of the Heisenberg group (cf. (0.17) for a definition of the unitary action associated with the translation above). So far as symplectic transformations Ψ are concerned, the relevant representation $\Psi \mapsto \tilde{\Psi}$ (actually a genuine representation of a two-fold covering of $Sp(n,\mathbb{R})$) had been built by A. Weil [W2]. He had actually not busied himself with the Weyl (or any symbolic) calculus, rather with understanding Siegel's theta functions (another account of that story was given later by Cartier [C3]). However, some physicists were aware of the *covariance formula*

$$\tilde{\Psi}\, Op(f)\, \tilde{\Psi}^{-1} = Op(f \circ \Psi^{-1}) \qquad (0.51)$$

often ascribed to Shale. That it took some time to be rediscovered by partial differential equations people is of course due to the fact that it works only with the Weyl calculus, not with the standard calculus.

The representation $\Psi \mapsto \tilde{\Psi}$ of (a two-fold covering of) $Sp(n,\mathbb{R})$ is called the *metaplectic* representation. Leray's lectures on Lagrangian Analysis [L1] and his account of Maslov's work brought it to the attention of analysts. It immediately became clear, then, that Beals' classes $S(\Phi,\varphi)$ should be redefined in some $Sp(n,\mathbb{R})$-invariant way, using of course the Weyl calculus. This was done independently by Hörmander [H4] and one of the present authors [U1, U3]. The basic concept in [H4] is that of a Riemannian structure on $\mathbb{R}^{2n}$; the one in [U3] is a family of harmonic oscillators. They amount to the same in the case when $\Phi\varphi = 1$ (or the $Sp(n,\mathbb{R})$-invariant analogue), which is the only one considered in [U3], contrary to [H4]. However, any class $S(\Phi,\varphi)$ would be contained in infinitely many classes $S(\varphi,\varphi^{-1})$, just as classes $S_{1,0}$ of classical symbols are contained in $S_{\delta,\delta}$ for all δ, $0 \leq \delta < 1$.

We shall now very briefly describe the pseudodifferential operator theory based on families of harmonic oscillators, since it is the one that leads to the introduction of coherent states (of Gaussian type) as a tool in pseudodifferential analysis. Coherent states of such a kind were discussed in 1978, independently, by Cordoba and Fefferman [CF] (who called them wave packets) and one of the authors [U2]. Let us describe the structure before telling the differences. For simplicity of notation, let us denote as z (resp. w) the current point (x,ξ) (resp. (y,η)) of $\mathbb{R}^{2n}$, it being understood, however, that in what follows z,w, ... really belongs to $\mathbb{R}^{2n}$, not $\mathbb{C}^n$: indeed, only the symplectic structure on $\mathbb{R}^{2n}$ is *fixed*, no complex structure is. Assume that, to each point $w \in \mathbb{R}^{2n}$, one has associated a certain Euclidean norm $\| \,\|_w$ on $\mathbb{R}^{2n}$, actually the image of the standard norm on $\mathbb{R}^{2n}$ under some symplectic transformation. One may wish to picture the structure as that of a family of "squeezed" balls B_w centered at all points of $\mathbb{R}^{2n}$; in the case when the map $w \mapsto \| \,\|_w$ is smooth, one, of course, has a Riemannian structure, which was Hörmander's concept [H4]. Any constant vector field D on $\mathbb{R}^{2n}$ can be called admissible at w if $\|D\|_w \leq 1$. Then, one can define symbols of weight one as those functions $f \in C^\infty(\mathbb{R}^{2n})$ such that, given j, there is a uniform bound for $|(D_1 \ldots D_j)f(w)|$ if $D_1, \ldots, D_j$ are admissible at w. Note that this is a straightforward generalization of Beals' definition of the classes $S(\Phi,\varphi)$. Given w and $w' \in \mathbb{R}^{2n}$, set, for any $z \in \mathbb{R}^{2n}$,

$$\|z\|^2_{w,w'} = 2 \inf\{\|z_1\|^2_w + \|z_2\|^2_{w'} : z_1 + z_2 = z\}. \tag{0.52}$$

Then, under Weyl's rule, symbols of weight one give rise to bounded operators on $L^2(\mathbb{R}^n)$ if the family $w \mapsto \| \,\|_w$ satisfies the requirement that

$$k(w,w') := (1 + \|w - w'\|^2_{w,w'})^{-N} \tag{0.53}$$

is, for large N, the kernel of a bounded operator on $L^2(\mathbb{R}^{2n})$. A stronger condition, introduced by Hörmander [H4], would be that the inequality

$$\|z\|_{w'} \leq C_1 \|z\|_w (1 + \|w - w'\|_w)^{N_1} \tag{0.54}$$

should hold for some pair of constants (C_1,N_1).

Now, as a generalization of (0.24), one can introduce, for every $w \in \mathbb{R}^{2n}$, the harmonic oscillator L_w centered at w, defined as

$$L_w = \mathrm{Op}(z \mapsto \pi \|z - w\|^2_w). \tag{0.55}$$

It has a discrete spectrum and has a complete family of eigenstates (φ^α_w), $\alpha \in \mathbb{N}^n$, that are Hermite functions. The basic one φ_w is the image under the Heisenberg map $\varphi \mapsto \varphi_z$ defined in (0.17), with $z = y + i\eta$ replaced by w, of some nonstandard Gaussian function $t \mapsto \exp(-\pi \langle At,t\rangle)$, where $A \in M_n(\mathbb{C})$ is its own transpose, and Re A is positive-definite; A depends on the symplectic norm $\| \,\|_w$, in a way more easily made explicit if one introduces Siegel's domain (cf. [U3]). Of course, with proper normalization, one may write $\|u\|^2 = \Sigma\, |(\varphi^\alpha_w \mid u)|^2$ for any *fixed* w if *all* α are allowed. One result of [U2] is that there is some *finite number* $k \in \mathbb{N}$ such that the family (φ^α_w) with $w \in \mathbb{R}^{2n}$ and $|\alpha| \leq k$ is a family of coherent states in

$L^2(\mathbb{R}^n)$ in the sense that (0.16) is valid. Cordoba and Fefferman [CF] concerned themselves only with the Riemannian structure associated with the class $S_{1,0}$ (i.e., symbols of classical type): also, they used only the Gaussian functions φ^0_w, but then, of necessity, only got (0.16) up to some error term bounded by $C\,\|u\|^2_{-1/2}$, where $\|\ \|_{-1/2}$ here denotes the usual Sobolev norm.

Finally, here is how all this is related to Theorem 0.27 as it is reproduced here. Under the assumption above [the weaker one, explained right after (0.53)], given any symbol f of weight one, any pair (α,β) of multi-indices $\in \mathbb{N}^n$, and any $N \in \mathbb{N}$, there exists $C > 0$ such that

$$|(\varphi^{\alpha}_{w'}|Op(f)\varphi^{\beta}_{w})| \leq C\,(1 + \|w - w'\|^2_{w,w'})^{-N} \tag{0.56}$$

for all $w,w' \in \mathbb{R}^{2n}$. This is a consequence of the results in [U3]. It would be a trivial matter to extend this to more general weights. The converse, however, cannot be true without any further assumption, and was proved by F. Bruyant [B8] under the hypothesis that, denoting as d the geodesic distance on $\mathbb{R}^{2n}$ associated with the given Riemannian structure, there exist $C > 0$ and $N > 0$ such that

$$C^{-1}\,(1 + d(w,w'))^{N^{-1}} \leq 1 + \|w - w'\|_w \leq C\,(1 + d(w,w'))^N \tag{0.57}$$

for all $w,w' \in \mathbb{R}^{2n}$.

From all that precedes we hope that the reader became convinced that, even when no neat group structure is present, families of coherent states are a highly adaptable tool for pseudodifferential analysis.

A semi-group generalization of the metaplectic representation was introduced by Howe [H6], and a report on many of the matters discussed in the present section was included in Folland [F2]. It would lead us astray from the topic in this section, namely pseudodifferential analysis on $\mathbb{R}^n$, to tell how all this led to quantization methods on phase-space. It is clear, however, that all the authors working in this field have been under the influence of representation theory, in particular under that of the metaplectic representation.

1

General Definition of the Fuchs Calculus

The Fuchs calculus is a symbolic calculus of operators on certain domains Λ of $\mathbb{R}^n$: both the internal geometry of Λ, described by a Riemannian structure, and the realization of Λ as an open subset of $\mathbb{R}^n$, play a role in the very definition of the calculus; far from being "less intrinsic," this dependence on the embedding in $\mathbb{R}^n$ makes it possible, for instance, to distinguish between $\mathbb{R}$ and a half-line, for which the calculi are totally different.

We shall assume in all that follows that Λ is an open subset of $\mathbb{R}^n$, provided with a Riemannian structure that makes it a globally symmetric space. Let us denote as S_z the geodesic symmetry around $z \in \Lambda$. Also assume that, given x and $y \in \Lambda$, there is a unique point $z \in \Lambda$, the geodesic middle of x and y, such that $y = S_z x$; let us denote it as $z = \text{mid}(x,y)$. Unless $\Lambda = \mathbb{R}^n$, the embedding $\Lambda \subset \mathbb{R}^n$ will not be isometric, i.e., the Riemannian structure on Λ is not the one induced from $\mathbb{R}^n$. Let dy denote the Lebesgue measure on $\mathbb{R}^n$. The standard inner product $\langle y,\eta \rangle$ on $\mathbb{R}^n$ identifies $\mathbb{R}^n$ with its linear dual space. It follows that the cotangent bundle $T^*\Lambda$ of Λ can be identified with $\Lambda \times \mathbb{R}^n$. On $T^*\Lambda$ we will always use the Lebesgue measure $dy d\eta$.

DEFINITION 1.1 Let dm be a measure on Λ invariant under all geodesic symmetries S_y, $y \in \Lambda$. Let

$$L^2(\Lambda) := L^2(\Lambda, dm) \tag{1.2}$$

denote the Hilbert space of all (classes of) complex-valued functions on Λ that are square-integrable with respect to dm. The scalar product on $L^2(\Lambda)$ is defined by

$$(v|u) := \int_{\Lambda} \overline{v}(t)u(t)\, dm(t). \tag{1.3}$$

Note that (1.3) is conjugate-linear in the first variable, a convention that we prefer for inner products in function spaces.

PROPOSITION 1.4
For any pair $(y,\eta) \in \Lambda \times \mathbb{R}^n$, *the formula*

$$(\sigma_{y,\eta}u)(t) := u(S_y t)\, e^{2\pi i \langle \eta, t - S_y t\rangle} \tag{1.5}$$

defines a unitary self-adjoint operator on $L^2(\Lambda)$. *Here* $u \in L^2(\Lambda)$ *and* $t \in \Lambda$.

PROOF Use the fact that S_y is measure preserving as well as involutive. Q.E.D.

DEFINITION 1.6 If f is a summable function on $\Lambda \times \mathbb{R}^n$, Op(f) shall be the (bounded) operator defined on $L^2(\Lambda)$ by

$$(v|\mathrm{Op}(f)u) := 2^n \int_{T^*\Lambda} f(y,\eta)\, (v|\sigma_{y,\eta}u)\, dy\, d\eta \tag{1.7}$$

for all $u,v \in L^2(\Lambda)$. We shall refer to f as the *active (Fuchs) symbol* of Op(f).

One may write formally

$$(\mathrm{Op}(f)u)(t) = 2^n \int_{T^*\Lambda} f(y,\eta)\, u(S_y t)\, e^{2\pi i\langle \eta, t - S_y t\rangle}\, dy\, d\eta \tag{1.8}$$

or, in short,

$$\mathrm{Op}(f) = 2^n \int_{T^*\Lambda} f(y,\eta)\, \sigma_{y,\eta}\, dy\, d\eta. \tag{1.9}$$

DEFINITION 1.10 If A is a trace-class operator on $L^2(\Lambda)$, the *passive (Fuchs) symbol* of A shall be the continuous function g = Symb(A) on $\Lambda \times \mathbb{R}^n$ defined by

$$g(y,\eta) = 2^n\, \mathrm{Tr}(A\, \sigma_{y,\eta}) \tag{1.11}$$

for all $(y,\eta) \in \Lambda \times \mathbb{R}^n$. Here Tr is the trace.

It is quite common that a natural quantization process leads one to introducing two species of symbols. We saw an example in Section 0 with the standard and antistandard symbols on $\mathbb{R}^n$. Many other instances have been met now, including the *contravariant* and *covariant* symbols in Berezin's calculus [B3, B4]. Sometimes, as shown (cf. Section 0) by the Weyl calculus on $\mathbb{R}^n$, one can make the two species meet halfway. This is not essential (it would be possible here, but would only make things more complicated at the present time). What is essential, however, and very far from being true, say, in the Berezin calculus, is that the two species of symbols should be related in such a way as to satisfy the same kinds of estimates characterizing symbol classes; that this is the case with the Fuchs calculus will be shown in Section 11.

PROPOSITION 1.12
The two maps Op *and* Symb *are formally adjoint to each other when viewed as maps between the space of square-summable functions on* $\Lambda \times \mathbb{R}^n$ *and the space of Hilbert-Schmidt operators on* $L^2(\Lambda)$.

PROOF Suppose $f \in L^1(\Lambda \times \mathbb{R}^n)$, A is trace-class, and g is the passive symbol of A. Then

$$\begin{aligned} \mathrm{Tr}(A\, \mathrm{Op}(f)) &= 2^n \int_{T^*\Lambda} f(y, \eta)\, \mathrm{Tr}(A\, \sigma_{y,\eta})\, dy\, d\eta \\ &= \int_{T^*\Lambda} f(y, \eta)\, g(y, \eta)\, dy\, d\eta. \end{aligned} \tag{1.13}$$

Since the adjoint $\mathrm{Op}(f)^*$ of Op(f) is just $\mathrm{Op}(\bar{f})$, where $\bar{f}$ is the complex-conjugate of f, the assertion follows. Q.E.D

PROPOSITION 1.14
Assume that the map

$$(s,t) \mapsto (\mathrm{mid}(s,t), s - t) \tag{1.15}$$

is a global diffeomorphism from $\Lambda \times \Lambda$ *onto* $\Lambda \times \mathbb{R}^n$. *Then the operator* Op(f) *with active symbol* $f \in L^1(\Lambda \times \mathbb{R}^n)$ *has the kernel*

$$k(s,t) = 2^n\, (\mathcal{F}_2^{-1} f)(\mathrm{mid}(s,t), s - t)\, \mathrm{Det}\left(\frac{\partial}{\partial t}\, \mathrm{mid}\,(s,t)\right) \tag{1.16}$$

with respect to Lebesgue measure. Here

$$(\mathcal{F}_2^{-1}f)(y,t) = \int_{\mathbb{R}^n} f(y,\eta)\, e^{2\pi i\langle \eta,t\rangle}\, d\eta \tag{1.17}$$

denotes the inverse Fourier transform of f with respect to the second variable.

PROOF Performing a change of variables in (1.8), one may write

$$(\mathrm{Op}(f)u)(s) = 2^n \int_{T^*\Lambda} f(\mathrm{mid}(s,t),\eta)\, u(t)\, e^{2\pi i\langle \eta, s-t\rangle} \cdot \mathrm{Det}\left(\frac{\partial}{\partial t}\mathrm{mid}(s,t)\right) dt\, d\eta$$

$$= 2^n \int_{\Lambda} (\mathcal{F}_2^{-1}f)(\mathrm{mid}(s,t), s-t)\, \mathrm{Det}\left(\frac{\partial}{\partial t}\mathrm{mid}(s,t)\right) u(t)\, dt.$$

Q.E.D.

Let us now turn to general examples of symmetric spaces Λ satisfying the assumption of Proposition 1.14.

Example 1.18
Let $\Lambda := \mathbb{R}^n$ *endowed with its standard euclidean structure. Then* Λ *is a symmetric space with geodesic symmetries given by*

$$S_y t = 2y - t \tag{1.19}$$

for all $y,t \in \mathbb{R}^n$. *Since the geodesic middle is just the euclidean midpoint* $mid(s,t) = (s+t)/2$, *the mapping* $(s,t) \mapsto ((s+t)/2, s-t)$ *defined in (1.15) is a diffeomorphism. Letting* $dm(y) = dy$ *be the Lebesgue measure on* $\mathbb{R}^n$ [*which is invariant under all symmetries (1.19)*], *we obtain the Hilbert space symmetries*

$$(\sigma_{y,\eta}u)(t) := u(2y-t)\, e^{4\pi i\langle \eta, t-y\rangle} \tag{1.20}$$

on $L^2(\mathbb{R}^n)$, *and the operator*

$$(\mathrm{Op}(f)u)(t) = 2^n \int_{\mathbb{R}^{2n}} f(y,\eta)\, u(2y-t)\, e^{4\pi i\langle \eta, t-y\rangle}\, dy\, d\eta$$

$$= \int_{\mathbb{R}^{2n}} f\left(\frac{s+t}{2}, \eta\right) u(s)\, e^{2\pi i\langle \eta, t-s\rangle}\, ds\, d\eta \tag{1.21}$$

on $L^2(\mathbb{R}^n)$, *with active Fuchs symbol f on* $T^*\Lambda = \mathbb{R}^{2n}$. *Thus the Fuchs calculus* *of Definition 1.6 is just the Weyl calculus [W3, H4, U1] for which the two species of symbols agree.*

Example 1.22
Let $\Lambda = \mathbb{R}_+ := \{y \in \mathbb{R} : y > 0\}$ be the open half-line endowed with the measure

$$dm(y) := \frac{dy}{y}. \tag{1.23}$$

Then the diffeomorphism $exp : \mathbb{R} \to \Lambda$ turns Λ into a Riemannian symmetric space with symmetries

$$S_y t = \frac{y^2}{t} \tag{1.24}$$

for all $y,t \in \mathbb{R}_+$. In particular,

$$S_e t = \frac{1}{t} \tag{1.25}$$

for the "base-point" $e := 1$ of Λ. For $s,t \in \Lambda$ the geodesic middle is given by $mid(s,t) = (st)^{1/2}$, and it is easy to show that the basic assumption of Proposition 1.14 is satisfied. Also, Λ is the basic example of a symmetric cone *(treated in Section 2).*

Example 1.26
Suppose $\Lambda_1 \subset \mathbb{R}^n$ and $\Lambda_2 \subset \mathbb{R}^m$ are symmetric spaces both satisfying the basic assumption of Proposition 1.14. Consider the product space $\Lambda := \Lambda_1 \times \Lambda_2 \subset \mathbb{R}^{n+m}$ endowed with its natural Riemannian structure. Then Λ is again a symmetric space, with symmetries given by

$$S_{y_1,y_2}(t_1, t_2) = (S_{y_1}t_1, S_{y_2}t_2) \tag{1.27}$$

for all y_j, $t_j \in \Lambda_j$ and $j \in \{1,2\}$. Clearly, Λ satisfies the basic assumption of Proposition 1.14 as well. Taking the product measure $dm(y_1,y_2) := dm_1(y_1)dm_2(y_2)$ on Λ, we obtain the Fuchs calculus on

$$L^2(\Lambda) = L^2(\Lambda_1)\hat{\otimes}\, L^2(\Lambda_2) \tag{1.28}$$

by tensoring the Fuchs calculi on Λ_1 and Λ_2, respectively. This applies in particular to the half-space

$$\mathbb{R}_+^{n+1} := \mathbb{R}_+ \times \mathbb{R}^n = \{(x_0, \ldots, x_n) \in \mathbb{R}^{n+1} : x_0 > 0\}. \tag{1.29}$$

To end this section let us see, in a formal sense, whether we can attribute to $A = \mathrm{Op}(f)$ a *standard symbol* in the sense that for some function $a(y,\eta)$ on $\Lambda \times \mathbb{R}^n$ we can also write

$$\begin{aligned}(Au)(s) &= \int_{T^*\Lambda} a(s, \xi) e^{2i\pi\langle \xi, s-t\rangle} u(t)\, d\xi\, dt \\ &= \int_{T^*\Lambda} (\mathcal{F}_2^{-1} a)(s, s-t)\, u(t)\, dt.\end{aligned} \tag{1.30}$$

It is clear that the required condition is

$$(\mathcal{F}_2^{-1} a)(s,s-t) = 2^n\, (\mathcal{F}_2^{-1} f)(\mathrm{mid}(s,t),s-t)\ \mathrm{Det}\left(\frac{\partial}{\partial t}\, \mathrm{mid}(s,t)\right), \tag{1.31}$$

a condition that only puts a restriction on the values of $(\mathcal{F}_2^{-1} a)(s,t)$ on the set where $s - t \in \Lambda$. We shall extend $\mathcal{F}_2^{-1} a$ by zero elsewhere. By (1.31) we obtain

$$\begin{aligned} &a(s, \xi) \\ &= 2^n \int_{T^*\Lambda} f(\mathrm{mid}(s, t), \eta) e^{2i\pi\langle \eta-\xi, s-t\rangle}\ \mathrm{Det}\left(\frac{\partial}{\partial t}\, \mathrm{mid}(s,t)\right) dt\, d\eta.\end{aligned} \tag{1.32}$$

To invert the correspondence $f \mapsto a$, we start from (1.31) rewritten as

$$(\mathcal{F}_2^{-1} f)(y, s - S_y s) = 2^{-n}\ \mathrm{Det}\left(\frac{\partial}{\partial y}\, (S_y s)\right) (\mathcal{F}_2^{-1} a)(s, s - S_y s). \tag{1.33}$$

As

$$f(y, \eta) = \int_{T^*\Lambda} (\mathcal{F}_2^{-1} f)(y, s - S_y s) e^{-2i\pi\langle \eta, s - S_y s\rangle}\ \mathrm{Det}\left(\frac{\partial}{\partial s}\, (s - S_y s)\right) ds, \tag{1.34}$$

we finally obtain

$$\begin{aligned} f(y, \eta) = 2^{-n} \int_{T^*\Lambda} & e^{2i\pi\langle \xi - \eta, s - S_y s\rangle} a(s, \xi) \cdot \mathrm{Det}\left(\frac{\partial}{\partial y}\, (S_y s)\right) \\ & \mathrm{Det}\left(\frac{\partial}{\partial s}\, (s - S_y s)\right) ds\, d\xi. \end{aligned} \tag{1.35}$$

2

The Geometry of Symmetric Cones

The present work is concerned with the case when the configuration space Λ discussed in Section 1 is a *symmetric cone.* In this section we give a unified description of all symmetric cones, using the theory of Jordan algebras [F1, FK2, BK] and illustrate the general theory by means of examples. With the exception of Proposition 2.88, this section is purely expository.

DEFINITION 2.1 A convex open cone $\Lambda \subset \mathbb{R}^n$ is called *symmetric* if it is *homogeneous* under its linear automorphism group

$$GL(\Lambda) := \{P \in GL(n,\mathbb{R}) : P\Lambda = \Lambda\}, \tag{2.2}$$

and it is *self-dual* in the sense that its closure $\bar{\Lambda}$ satisfies

$$\bar{\Lambda} = \{y \in \mathbb{R}^n: \langle y,x \rangle \geq 0 \; \forall x \in \bar{\Lambda}\} \tag{2.3}$$

This implies in particular that Λ is proper, i.e., does not contain any line through the origin. The group $GL(\Lambda)$ is a closed subgroup of $GL(n,\mathbb{R})$ and is therefore a real Lie group. The self-duality condition on Λ implies

$$P \in GL(\Lambda) \Rightarrow P' \in GL(\Lambda), \tag{2.4}$$

where P' is the transpose of $P \in GL(n,\mathbb{R})$. It is also known [cf. (2.57)] that

$$P \in GL(\Lambda), P > 0 \Rightarrow P^{1/2} \in GL(\Lambda),$$

where $P > 0$ means positive definite. Define

$$O(\Lambda) := GL(\Lambda) \cap O(n). \tag{2.5}$$

By [F1; Theorem 1.4.3], (cf. also [FK2]), $O(\Lambda)$ is a maximal compact subgroup of $GL(\Lambda)$, and there exists a "base point" $e \in \Lambda$ such that

$$O(\Lambda) = \{P \in GL(\Lambda): Pe = e\} \tag{2.6}$$

is the stabilizer of e.

PROPOSITION 2.7
For any point $y \in \Lambda$, *choose* $P \in GL(\Lambda)$ *such that* $Pe = y$. *Then the matrix*

$$P_y := PP' \in GL(\Lambda) \tag{2.8}$$

depends only on y. *Moreover, we have* $P_y^{1/2}\, e = y$ *and*

$$P_{Qy} = QP_yQ' \tag{2.9}$$

whenever $Q \in GL(\Lambda)$.

PROOF Suppose $P, Q \in GL(\Lambda)$ satisfy $Pe = Qe = y$. Then $Q^{-1}P \in O(\Lambda) \subset O(n)$, and $P = Q(Q^{-1}P)$ satisfies

$$PP' = Q(Q^{-1}P)(Q^{-1}P)'Q' = QQ',$$

showing that P_y depends only on y. As

$$(P^{-1}P_y^{1/2})(P^{-1}P_y^{1/2})' = P^{-1}P_yP'^{-1} = I \qquad \text{(identity)},$$

the matrix $P^{-1}P_y^{1/2}$ belongs to $O(\Lambda)$, so that $P_y^{1/2}\, e = Pe = y$. The last formula follows from the definition. Q.E.D.

The cone Λ has a unique $GL(\Lambda)$-invariant Riemannian structure such that, for the canonical identification of $T_e(\Lambda)$ with $\mathbb{R}^n$, the norm on $T_e(\Lambda)$ is given by the standard norm. As recalled next, Λ is a *globally symmetric space* for this structure—in particular, one may define dm (cf. Definition 1.1) by means of the Riemannian structure, since the latter one is invariant under all geodesic symmetries.

PROPOSITION. 2.10
The mapping

$$y \mapsto Sy := P_y^{-1/2}e \qquad (y \in \Lambda) \tag{2.11}$$

is an involutive isometry of Λ *preserving* e *(called the symmetry of* Λ *around* e *). It satisfies*

$$SP^{-1} = P'S \tag{2.12}$$

for all $P \in GL(\Lambda)$. *If* $x = Pe$, *the symmetry* S_x *of* Λ *around* x *is given by*

$$S_x = PSP^{-1} = PP'S. \tag{2.13}$$

PROOF Let $y = Pe$ for some $P \in GL(\Lambda)$. Then the point $P'^{-1}e \in \Lambda$ depends only on y since

$$Qe = Pe \Rightarrow Q^{-1}P \in O(\Lambda) \Rightarrow P'Q'^{-1} \in O(\Lambda) \Rightarrow Q'^{-1}e = P'^{-1}e$$

for any $Q \in GL(\Lambda)$. Specializing to $P := P_y^{1/2}$ we see that

$$S(Pe) = P'^{-1}e$$

for every $P \in GL(\Lambda)$. This implies that S is involutive. Now let d be the geodesic distance on Λ. Since d is invariant under $GL(\Lambda)$, we have

$$d(e, Pe) = d(e, P_y^{1/2}e) = d(P_y^{-1/2}e, e) = d(P'^{-1}e, e),$$

and

$$d(Qe,Pe) = d(e,Q^{-1}Pe) = d(Q'P'^{-1}e,e) = d(P'^{-1}e,Q'^{-1}e)$$

for all $P,Q \in GL(\Lambda)$. Thus S is an isometry of the metric space (Λ,d). If $x = Qe$, then $SP^{-1}x = SP^{-1}Qe = P'Q'^{-1}e = P'Sx$. This implies (2.12) and hence (2.13). Q.E.D.

Example 2.14
Consider the space $R^{r\times r}$ *of all real* $(r \times r)$*-matrices*

$$x = (x_{jk})_{1\le j,k\le r}$$

and its subspace

$$\mathcal{H}_r(\mathbb{R}) := \{x \in \mathbb{R}^{r\times r} : x' = x\} \tag{2.15}$$

of all symmetric matrices. Via the scalar product

$$\langle x | y\rangle := \mathrm{Tr}(xy) = \sum_{j,k} x_{jk}y_{jk}, \tag{2.16}$$

$\mathcal{H}_r(\mathbb{R})$ can be identified with the canonical euclidean space of dimension $n = r(r+1)/2$, by means of the coordinates x_{jj} $(1 \leq j \leq r)$ and $\sqrt{2}\, x_{jk}$ $(1 \leq j < k \leq r)$. One can show that the convex open cone

$$\Lambda_{\mathbb{R}} := \{y \in \mathcal{H}_r(\mathbb{R}) : y > 0\} \tag{2.17}$$

of all positive definite symmetric matrices is self-dual with respect to (2.16) and is also homogeneous under the group $GL(\Lambda_{\mathbb{R}}) \subset GL(n,\mathbb{R})$ consisting of all transformations

$$P_a y := aya' \ (y \in \Lambda_{\mathbb{R}}), \tag{2.18}$$

where $a \in GL(r,\mathbb{R})$. Thus $\Lambda_{\mathbb{R}}$ is a symmetric cone. For $r = 1$, $\Lambda_{\mathbb{R}} = \mathbb{R}_+$ is the half-line. The transformations (2.18) have the transpose

$$P_a' = P_{a'}$$

with respect to (2.16), and we have

$$O(\Lambda_{\mathbb{R}}) = \{P_a : a \in O(r)\} \tag{2.19}$$

when choosing the base point $e := I_r$ (identity matrix) in $\Lambda_{\mathbb{R}}$. In the special case $y \in \Lambda_{\mathbb{R}} \subset GL(n\mathbb{R})$, the mapping

$$P_y x = yxy$$

agrees with the endomorphism P_y defined in (2.8) since clearly $P_{y^{1/2}} \in GL(\Lambda_{\mathbb{R}})$ is self-adjoint and satisfies $P_{y^{1/2}} e = y$ and $(P_{y^{1/2}})^2 = P_y$. For the symmetry S of $\Lambda_{\mathbb{R}}$ around e this implies

$$S(y) = y^{-1} \tag{2.20}$$

for all $y \in \Lambda_{\mathbb{R}}$.

Example 2.21

Generalizing Example 2.14, let $\mathbb{K}$ *denote one of the real division algebras* $\mathbb{R}$ *(real numbers),* $\mathbb{C}$ *(complex numbers), or* $\mathbb{H}$ *(quaternions) endowed with its natural involution* $\alpha \mapsto \bar{\alpha}$. *Consider the space* $\mathbb{K}^{r\times r}$ *of all* $(r\times r)$*-matrices*

$$x = (x_{jk})_{1\le j,k\le r}$$

with entries $x_{jk} \in \mathbb{K}$, *and its subspace*

$$\mathcal{H}_r(\mathbb{K}) = \{x \in \mathbb{K}^{r\times r} : x^* = x\} \tag{2.22}$$

of all self-adjoint matrices (satisfying $\bar{x}_{jk} = x_{kj}$ *for all* j,k*). This is a euclidean space of dimension*

$$n = r + \frac{a}{2} r(r-1) \tag{2.23}$$

where $a := \dim_{\mathbb{R}}^{\mathbb{K}} \in \{1,2,4\}$. One can show that the convex open cone

$$\Lambda_{\mathbb{K}} := \{y \in \mathcal{H}_r(\mathbb{K}) : y > 0\} \tag{2.24}$$

of all positive definite self-adjoint matrices over $\mathbb{K}$ *is self-dual and is also homogeneous under the group* $GL(\Lambda_{\mathbb{K}}) \subset GL(n,\mathbb{R})$ *consisting of all transformations*

$$P_a y = aya^* \quad (y \in \Lambda_{\mathbb{K}}), \tag{2.25}$$

where $a \in GL(r,\mathbb{K})$ *is an invertible* $(r\times r)$*-matrix with entries in* $\mathbb{K}$. *Here* $(a_{jk})^* = (\bar{a}_{kj})$ *is the adjoint matrix. Thus* $\Lambda_{\mathbb{K}}$ *is a symmetric cone. For* $r = 1$, $\Lambda_{\mathbb{K}} = \mathbb{R}_+$ *is the half-line. As in Example 2.14, we have the relations*

$$P'_a = P_{a^*} \qquad (a \in GL(r, \mathbb{K})),$$

$$P_y x = yxy \qquad (y \in \Lambda_{\mathbb{K}})$$

and

$$O(\Lambda_{\mathbb{K}}) = \{P_a : a \in U(r,\mathbb{K})\} \tag{2.26}$$

(unitary group over $\mathbb{K}$*) if the base point* $e \in \Lambda_{\mathbb{K}}$ *is chosen as the identity matrix of rank* r *over* $\mathbb{K}$.

Example 2.27
The second fundamental class of symmetric cones, besides the matrix cones Λ_K *described in Examples 2.14 and 2.21, are the (solid) light cones. Put* $n = d + 1$ *and denote points in* $\mathbb{R}^n$ as

$$x = (x_0, x_1, \dots, x_d) = (x_0, \mathbf{x}),$$

a convention standard in relativity. Consider the quadratic form

$$\Delta(x) := x_0^2 - \mathbf{x} \cdot \mathbf{x}, \tag{2.28}$$

where $\cdot$ *denotes the dot product in* $\mathbb{R}^d$. *Then*

$$\Lambda_n := \{x \in \mathbb{R}^n \colon \Delta(x) > 0, x_0 > 0\} \tag{2.29}$$

is the solid forward light cone of dimension n. When $n = 1$ *or 2,* Λ_n *coincides with* $\mathbb{R}_+$ *or* $\mathbb{R}_+ \times \mathbb{R}_+$, *respectively. For* $n \geq 3$, Λ_n *is an irreducible symmetric cone, i.e.,* Λ_n *is not a direct product of lower-dimensional cones. Let O(1,d) be the Lorentz group of the quadratic form (2.28). Then we have*

$$GL(\Lambda) = \{\lambda P \colon \lambda > 0, P \in O(1,d) \text{ orthochronous}\}: \tag{2.30}$$

recall that a Lorentz transformation is orthochronous if it does not reverse time. Choosing the base point $e := (1,\mathbf{0})$ *we see that*

$$O(\Lambda_n) \cong O(d) \tag{2.31}$$

consists of all purely spatial isometries.

In order to study the geometric properties of symmetric cones in more detail (necessary for estimates concerning the Fuchs calculus), we will now describe in some detail how a *Jordan algebraic structure* on $\mathbb{R}^n$ is associated with a symmetric cone Λ. By definition, a bilinear product

$$(x,y) \mapsto x \circ y \tag{2.32}$$

on $\mathbb{R}^n$ defines a *Jordan algebra* if the following identities hold:

$$x \circ y = y \circ x \qquad \text{(commutativity)}, \tag{2.33}$$

$$x^2 \circ (x \circ y) = x \circ (x^2 \circ y) \qquad \text{(Jordan identity)}. \tag{2.34}$$

Here we put $x^2 := x \circ x$. Historically [JNW] these axioms were derived algebraically, starting from abstract properties of the so-called *anti-commutator*

$$x \circ y := (xy + yx)/2 \tag{2.35}$$

of an associative product. (Not every Jordan algebra can be realized in this way, however). Following [BK, P2], we will instead use a geometric-analytic approach towards Jordan algebras, starting directly from the basic geometric properties of a given symmetric cone Λ (cf. Definition 2.1). In the course of the argument, one obtains not only the Jordan algebra product, but also the "quadratic representation" operators P_y [cf. (2.8)], the $GL(\Lambda)$-invariant Riemannian structure, and the symmetry S [cf. (2.11)] playing the role of an "inversion" map.

For a (possibly vector valued) smooth function f on Λ, let $\partial_x f$ denote the derivative at $x \in \Lambda$, regarded as a linear mapping. Then

$$\partial_x^u f := (\partial_x f)u \tag{2.36}$$

is the directional derivative in direction $u \in \mathbb{R}^n = T_x(\Lambda)$ (tangent space). Higher-order derivatives will be written in a slightly different way, for example $(\partial_x^2 f)(u,v)$ shall denote the second derivative evaluated at $u,v \in \mathbb{R}^n$. Now consider the (absolutely convergent) integral

$$I(x) := \int_\Lambda e^{-\langle x,y\rangle}\, dy, \tag{2.37}$$

which defines a (strictly positive) function of $x \in \Lambda$. The inverse

$$\omega(x) := I(x)^{-1} = \left(\int_\Lambda e^{-\langle x,y\rangle}\, dy\right)^{-1} \tag{2.38}$$

is called the *norm function* of Λ. It satisfies $\omega(\lambda x) = \lambda^n \omega(x)$ for $\lambda > 0$ and more generally

$$\omega(Px) = |\mathrm{Det}\, P| \cdot \omega(x) \tag{2.39}$$

for all $x \in \Lambda$ and $P \in GL(\Lambda)$, since P and its adjoint $P' \in GL(\Lambda)$ (with respect to $\langle x,y\rangle$) have the same determinant. Taking derivatives

$$\partial_x^u I = \int_\Lambda \partial_x^u e^{-\langle x,y\rangle}\, dy = -\int_\Lambda \langle u, y\rangle e^{-\langle x,y\rangle}\, dy,$$

$$(\partial_x^2 I)(u, v) = \int_\Lambda \langle u, y\rangle\langle v, y\rangle e^{-\langle x,y\rangle} dy$$

and

$$\partial_x^2(\log I)(u, v) = \partial_x^u \frac{1}{I(x)}\,(\partial_x^v I) = \frac{1}{I(x)^2}\,(I(x)\cdot \partial_x^u\partial_x^v I - \partial_x^u I\cdot \partial_x^v I)$$

it follows that

$$I(x)^2\cdot \partial_x^2(\log I)(u, v)$$

$$= \int_\Lambda dy_1 \int_\Lambda dy_2\, e^{-\langle x, y_1+y_2\rangle}[\langle u, y_1\rangle\langle v, y_1\rangle - \langle u, y_1\rangle\langle v, y_2\rangle].$$

Exchanging y_1 and y_2, we obtain

$$2\, I(x)^2\cdot \partial_x^2(\log I)(u, v) = -\frac{2}{\omega(x)^2}\,\partial_x^2(\log \omega)(u, v)$$

$$= \int_\Lambda dy_1 \int_\Lambda dy_2\, e^{-\langle x, y_1+y_2\rangle}\langle u, y_1 - y_2\rangle\, \langle v, y_1 - y_2\rangle. \tag{2.40}$$

Since this bilinear form is clearly positive definite, it follows that there exists a positive definite matrix P_x (with respect to $\langle,\rangle$) such that

$$-\partial_x^2(\log \omega)(u, v) = \langle P_x^{-1} u, v\rangle \tag{2.41}$$

for all $u,v \in \mathbb{R}^n$. For every $P \in GL(\Lambda)$, the chain rule and (2.39) imply

$$-\langle P_x^{-1}u, v\rangle = \partial_x^2(\log \omega)(u, v) = \partial_x^2(\log \omega \circ P)(u, v)$$
$$= \partial_{Px}^2(\log \omega)(Pu, Pv) = -\langle P_{Px}^{-1} Pu, Pv\rangle,$$

hence

$$P_{Px} = P\, P_x\, P', \tag{2.42}$$

a formula that should be compared with (2.9). Since (2.40) depends real-analytically on $x \in \Lambda$, (2.41) defines the (essentially unique) $GL(\Lambda)$-invariant Riemannian metric on Λ. Now define a real-analytic mapping $S : \Lambda \to \mathbb{R}^n$ by putting

$$\langle Sx, u \rangle := \partial_x^u \log \omega = \frac{1}{\omega(x)} \partial_x^u \omega. \tag{2.43}$$

This definition is a Jordan-theoretic "Cramer's rule," since Sx will turn out to be the algebraic inverse of x, and ω is related to a generalized determinant function. We have $S(\lambda x) = \lambda^{-1}\, Sx$ for $\lambda > 0$. Also note that $\partial_x^x\, I < 0$, which implies $\partial_x^x\, \omega > 0$, therefore $\langle Sx,x \rangle > 0$ for all $x \in \Lambda$.

PROPOSITION 2.44
S is an involutive diffeomorphism from Λ onto itself.

PROOF For any $x \in \Lambda$, consider the "norm-hypersurface"

$$\Lambda_x := \{z \in \Lambda : \omega(x) = \omega(z)\}. \tag{2.45}$$

For $y \in \mathbb{R}^n \setminus \{0\}$, let $y^\perp$ denote the orthogonal complement. We prove that the assumption $\langle y,x \rangle > 0$, together with

$$\langle y, z - x \rangle \geq 0 \text{ for all } z \in \Lambda_x \tag{2.46}$$

implies $y \in \bar{\Lambda}$ (closure of Λ). Geometrically, this means that if Λ_x lies on a certain side of the affine hyperplane $x + y^\perp$ then $y \in \bar{\Lambda}$. To prove (2.46), let $b \in \Lambda$. Since $z := (\omega(x)/\omega(b))^{1/n}\, b \in \Lambda_x$, we have

$$\langle y,b \rangle\, \omega(x)^{1/n} \geq \langle y,x \rangle\, \omega(b)^{1/n}. \tag{2.47}$$

By continuity, this inequality holds for all $b \in \bar{\Lambda}$, which gives $y \in \bar{\Lambda}$ since $\bar{\Lambda}$ is self-dual. This proves the implication above. Now let $x,z \in \Lambda$. By Taylor's formula, there exists $w \in [x,z] \subset \Lambda$ (line segment) such that

$$\begin{aligned} \log \omega(z) - \log \omega(x) &= \partial_x^{z-x} \log \omega + \frac{1}{2}\, \partial_w^2(\log \omega)(z - x, z - x) \\ &= \langle Sx, z - x \rangle - \frac{1}{2}\, \langle P_w^{-1}(z - x), z - x \rangle. \end{aligned}$$

Since P_w is positive definite, this implies

$$\log \omega(z) - \log \omega(x) \le \langle Sx, z - x\rangle. \tag{2.48}$$

Applying (2.46) together with the fact that $\langle Sx,x\rangle > 0$, we obtain $Sx \in \bar{\Lambda}$. Now we have

$$\langle \partial_x^u S,v\rangle = \partial_x^u \langle Sx,v\rangle = \partial_x^2 (\log \omega)(u,v) = -\langle P_x^{-1}u,v\rangle$$

for all $u,v \in \mathbb{R}^n$. Therefore

$$\partial_x S = -P_x^{-1} \tag{2.49}$$

for all $x \in \Lambda$. Since P_x is positive definite and hence invertible, the mapping S is locally bianalytic and therefore open. In particular, $S(\Lambda) \subset \Lambda$ since Λ is convex. The function $\varphi(x) := \omega(Sx)^{-1}$ is therefore well-defined on Λ. Now $\langle Sx,u\rangle = \partial_x^u \log \omega = \partial_x^u (\log \omega \circ P) = \partial_{Px}^{Pu} \log \omega = \langle S\,Px,Pu\rangle = \langle P'\,S\,Px,u\rangle$ for all $P \in GL(\Lambda)$. Therefore

$$S\,P = (P')^{-1}\,S, \tag{2.50}$$

and $\varphi(Px) = \omega(SPx)^{-1} = \omega((P')^{-1}\,Sx)^{-1} = [\,|\mathrm{Det}\,P|^{-1} \cdot \omega(Sx)]^{-1} = |\mathrm{Det}\,P| \cdot \varphi(x)$. Since Λ is homogeneous, this implies that φ is a constant multiple of ω, so that

$$\omega(x)\,\omega(Sx) = \text{const.} \tag{2.51}$$

for all $x \in \lambda$. Since $S(\lambda x) = \lambda^{-1}\,Sx$ for $\Lambda > 0$, Euler's relation gives

$$P_x^{-1}\,x = -\partial_x^x\,S = Sx. \tag{2.52}$$

It follows that

$$\begin{aligned} 0 = \partial_x^u \log[\omega(x)\omega(Sx)] &= \partial_x^u \log \omega + \partial_{Sx}(\log \omega)(\partial_x^u S) \\ &= \langle Sx, u\rangle + \langle S\,Sx, \partial_x^u S\rangle \\ &= \langle P_x^{-1}x, u\rangle - \langle S\,Sx, P_x^{-1}u\rangle. \end{aligned}$$

Therefore, $P_x^{-1}\,x = (P_x^{-1})'\,S\,Sx = P_x^{-1}\,S\,Sx$, and hence, $S\,Sx = x$. This shows that S is involutive and a bijection from Λ onto itself. Q.E.D.

Since S is involutive, we have

$$u = \partial_x^u x = \partial_x^u S\,Sx = (\partial_{Sx}S)(\partial_x^u S) = -P_{Sx}^{-1}\partial_x^u S = P_{Sx}^{-1}P_x^{-1}u,$$

which gives for all $x \in \Lambda$

$$P_{Sx} = P_x^{-1}. \tag{2.53}$$

PROPOSITION 2.54
S has a fixed point $e \in \Lambda$.

PROOF For every $x \in \Lambda$, (2.48) implies that Λ_x is on one side of $x + (Sx)^\perp$. Conversely, let $y \in \mathbb{R}^n \setminus \{0\}$ satisfy $\langle y,x\rangle > 0$ as well as (2.46). For $u \in y^\perp$ and $\lambda > 0$ small, we have $x \pm \lambda u \in \Lambda$ and (2.47) implies

$$\langle y, x\rangle = \langle y, x \pm \lambda u\rangle \geq \langle y, x\rangle(\omega(x \pm \lambda u)/\omega(x))^{1/n}.$$

Then $y \in \bar{\Lambda}$, hence $\omega(x \pm \lambda u) \leq \omega(x)$. It follows that

$$\pm\langle Sx, u\rangle = \partial_x^{\pm u} \log \omega = \lim_{\lambda\downarrow 0} \frac{\log \omega(x \pm \lambda u) - \log \omega(x)}{\lambda} \leq 0.$$

Therefore $\langle Sx,y^\perp\rangle = 0$ hence $Sx = \alpha y$, with $\alpha > 0$ since $Sx \in \Lambda$. Thus,

$$y = \alpha^{-1}(Sx) = S(\alpha x) \tag{2.55}$$

is uniquely determined (up to a positive factor). Now choose $x \in \Lambda$ with $\omega(x) = 1$. Since Λ_x has unique supporting planes (in the sense of the Hahn-Banach Theorem) there exists a (unique) $e \in \Lambda_x$ satisfying

$$\langle e,e\rangle = \inf \{\langle z,z\rangle : z \in \Lambda_x\}. \tag{2.56}$$

Using a Lagrange multiplier λ, one obtains for all $u \in \mathbb{R}^n$

$$0 = \partial_e^u(\langle z,z\rangle - \lambda\,(\omega(z) - 1)) = 2\,\langle e,u\rangle - 2\lambda\,\omega(e)\,\partial_e^u\omega.$$

Therefore,

$$\langle Se,u\rangle = \partial_e^u \log \omega = \omega(e)^{-1}\,\partial_e^u\,\omega = \partial_e^u\,\omega = \langle e,u\rangle/\lambda$$

showing that $\lambda > 0$ and $Se = \lambda^{-1}e$. By (2.51), we have $\lambda = 1$, so that e is a fixed point of S. Q.E.D.

Combining (2.53) and Proposition 2.54 we see that $P_e^2 = \mathrm{Id}$. Since P_e is positive definite, it follows that $P_e = \mathrm{Id}$, i.e., the invariant Riemannian metric coincides with the given inner product $\langle u,v\rangle$ at e. Since Λ is homogeneous, for every $x \in \Lambda$ there exists $P \in GL(\Lambda)$ satisfying $Pe = x$. By (2.42), this implies

$$P_x = P\, P_e\, P' \in GL(\Lambda). \tag{2.57}$$

Specializing $P = P_y$ in (2.42) yields the *fundamental formula*

$$P_{P_y x} = P_y\, P_x\, P_y \tag{2.58}$$

for all $x,y \in \Lambda$. Since $\langle \, , \, \rangle$ is a scalar product, one may define an algebra structure (i.e., a bilinear product) by putting for all $u,v,w \in \mathbb{R}^n$

$$\langle u \circ v, w \rangle := \frac{1}{2}\, \partial_e^3 (\log \omega)\, (u,v,w). \tag{2.59}$$

Since this defines a symmetric trilinear form, it follows that the product $u \circ v$ defined via (2.59) is commutative [i.e., (2.33) holds] and satisfies the associativity condition

$$\langle u \circ v, w \rangle = \langle u, v \circ w \rangle \tag{2.60}$$

with respect to the original inner product. (However, in general, $u \circ v$ is not associative in the algebraic sense). The identity (2.60) can also be expressed by saying that the *multiplication operators*

$$M_u\, v := u \circ v \tag{2.61}$$

on $\mathbb{R}^n$ are self-adjoint: $M_u' = M_u$. By definition, we have

$$2\, \langle M_u\, v, w \rangle = \partial_e^u\, [\partial_x^2 (\log \omega)\, (v,w)] = -\, \partial_e^u\, \langle P_x^{-1}\, v, w \rangle = -\, \langle (\partial_e^u P_x^{-1})\, v,w \rangle$$

so that (using $P_e = \mathrm{Id}$)

$$2\, M_u = -\partial_e^u P_x^{-1} = \partial_e^u P_x. \tag{2.62}$$

Applying (2.49) and (2.52) we obtain

$$-P_x^{-1}\, u = \partial_x^u\, S = \partial_x^u\, (P_x^{-1}\, x) = (\partial_x^u\, P_x^{-1})\, x + P_x^{-1}\, u,$$

therefore, $(\partial_x^u P_x^{-1})\, x = -2\, P_x^{-1} u$. Specializing to $x = e$, this yields

$$2u = 2P_e^{-1}\, u = -(\partial_e^u\, P_x^{-1})\, e = 2M_u e$$

showing that e is the unit element of $(\mathbb{R}^n, \circ)$. In particular, e is the *only* fixed point of S in Λ. Since

$$\omega(P_x y) = \text{Det}\, P_x \cdot \omega(y) \tag{2.63}$$

for all $x,y \in \Lambda$, we obtain by differentiation with respect to x

$$\begin{aligned} \partial_e^u \log \text{Det}\, P_x &= \partial_e^u (\log \omega)(P_x\, y) \\ &= \partial_y (\log \omega)(2u \circ y) \\ &= \langle Sy, 2u \circ y \rangle \\ &= 2\, \langle Sy \circ y, u \rangle. \end{aligned}$$

Since this must be independent of y, we obtain

$$Sy \circ y = Se \circ e = e \tag{2.64}$$

for all $y \in \Lambda$. (This indicates that $Sy = y^{-1}$ is the "inverse" of y in the Jordan-theoretic sense [BK]).

PROPOSITION 2.65
The multiplication (2.59) *defines a Jordan algebra on* $\mathbb{R}^n$, *and we have the "quadratic representation" formula*

$$P_x = 2M_x^2 - M_{x^2} \tag{2.66}$$

for all $x \in \mathbb{R}^n$.

PROOF For every $u \in \mathbb{R}^n$, (2.64) and the product rule imply

$$0 = \partial_x^u(x \circ Sx) = (\partial_x^u x) \circ Sx + x \circ (\partial_x^u S) = u \circ Sx - x \circ P_x^{-1} u$$

according to (2.49). In operator terms, this gives

$$M_{Sx} = M_x\, P_x^{-1}. \tag{2.67}$$

Since these operators are all self-adjoint, they commute. Hence

$$x^2 = M_x\, x = M_{Sx}\, P_x\, x = P_x\, M_{Sx}\, x = P_x\, (Sx \circ x) = P_x\, e$$

by (2.64). This implies, via (2.67), $x^2 \circ Sx = M_{Sx}\, P_x\, e = M_x\, e = x$ and differentiation yields

$$u = \partial_x^u\, (x^2 \circ Sx) = (\partial_x^u x^2) \circ Sx + x^2 \circ \partial_x^u S = 2\,(u \circ x) \circ Sx - x^2 \circ P_x^{-1} u.$$

Hence $\text{Id} = 2\, M_{Sx}\, M_x - M_{x^2}\, P_x^{-1}$ or, equivalently,

$$P_x = 2M_{Sx} M_x P_x - M_{x^2} = 2M_x^2 - M_{x^2}.$$

Since M_x and P_x commute, the Jordan identity (2.34) follows. Also, P_x is a quadratic polynomial in x and (2.66) holds for all $x \in \mathbb{R}^n$. Q.E.D.

Let $\mathfrak{gl}(\Lambda) \subset \mathfrak{gl}(n,\mathbb{R})$ denote the Lie algebra of the linear Lie group $GL(\Lambda)$ and consider the Lie algebra

$$o(\Lambda) := \{A \in \mathfrak{gl}(\Lambda) : A' + A = 0\} \tag{2.68}$$

of the closed subgroup $O(\Lambda)$. Putting

$$p := \{A \in \mathfrak{gl}(\Lambda) : A' = A\}, \tag{2.69}$$

we have the Cartan decomposition

$$\mathfrak{gl}(\Lambda) = o(\Lambda) \oplus p. \tag{2.70}$$

Using the exponential map $\exp : \mathbb{R}^n \to \Lambda$ defined in terms of the Jordan theoretic powers [cf. (2.87)], one can show that

$$\exp(2 M_x) = P_{\exp(x)} \tag{2.71}$$

for all $x \in \mathbb{R}^n$ [BK]. Therefore $M_x \in p$ and, in fact,

$$p = \{M_x : x \in \mathbb{R}^n\}. \tag{2.72}$$

This relationship describes the Jordan multiplication operators and hence the product directly in terms of the group $GL(\Lambda)$ and its Lie algebra. The inversion map

$$x \mapsto x^{-1} = Sx \tag{2.73}$$

in a Jordan algebra gives rise to two fundamental invariants. Since (2.73) is a rational mapping from (a dense subset of) $\mathbb{R}^n$ into $\mathbb{R}^n$ we can write

$$x^{-1} = \frac{\nabla(x)}{\Delta(x)} \tag{2.74}$$

where $\nabla : \mathbb{R}^n \to \mathbb{R}^n$ and $\Delta : \mathbb{R}^n \to \mathbb{R}$ are polynomials with no common scalar factors. One can show that, after normalizing $\Delta(e) = 1$, the expression (2.74) is unique. The polynomial Δ is called the (Jordan algebraic) *determinant* associated with Λ. The degree r of Δ is called the *rank* of the Jordan algebra. If we let λ be a scalar indeterminate, the expansion

$$\begin{aligned}\Delta(\lambda e - x) &= \lambda^r - a_1(x)\lambda^{r-1} + a_2(x)\lambda^{r-2} - \cdots + (-1)^r a_r(x) \\ &= \sum_{i=0}^{r} (-1)^i \lambda^{r-i} a_i(x) \end{aligned} \tag{2.75}$$

gives rise to new "invariant" polynomials $a_i : \mathbb{R}^n \to \mathbb{R}$ of degree i. We have $a_0 = 1$ and $a_r(x) = \Delta(x)$. The coefficient

$$\mathrm{tr}(x) := a_1(x) \tag{2.76}$$

is called the *trace* of x. One can show [F1; Proposition III.1.4] (cf. also [FK2]) that the $O(\Lambda)$-invariant bilinear form

$$(x,y) \mapsto \mathrm{tr}(x \circ y) \tag{2.77}$$

on $\mathbb{R}^n$ is positive definite and satisfies the self-duality condition (2.3). Define

$$|x| := \mathrm{tr}(x^2)^{1/2}. \tag{2.78}$$

For "irreducible" cones, the norm function $\omega(x)$ defined in (2.38) is the (n/r)-th power of $\Delta(x)$. Note that n/r need not be an integer (cf. Example 2.27). One can show [S5; §1] that the polynomial $\nabla(x)$ defined in (2.74) coincides with the gradient

$$\nabla(x) = \mathrm{grad}_x \Delta(x) \tag{2.79}$$

of $\Delta(x)$ with respect to (2.77). In the irreducible case, any two $O(\Lambda)$-invariant inner products are proportional, so that (2.77) coincides with the original scalar product on $\mathbb{R}^n$ up to normalization. At some points, it would seem more natural to use (2.77), but we stick with the original definition of $<x,y>$: this is, in any case, harmless since we shall be mostly interested in inequalities.

Example 2.80

The matrix cones $\Lambda_{\mathbb{K}}$ *(*$\mathbb{K} = \mathbb{R}, \mathbb{C}, \mathbb{H}$*) of size* $r \times r$ *have the linear automorphism group*

$$GL(\Lambda_{\mathbb{K}}) = \{P_a : a \in GL(r,\mathbb{K})\}$$

with Lie algebra

$$\mathfrak{gl}(\Lambda_{\mathbb{K}}) = \{M_a : a \in \mathfrak{gl}(r,\mathbb{K})\} \cong \mathfrak{gl}(r,\mathbb{K}).$$

Here we define

$$M_a x := \frac{1}{2}(ax + xa^*)$$

for all $a \in \mathfrak{gl}(r,\mathbb{K})$ *and* $x \in \mathcal{H}_r(\mathbb{K})$. *For all* $x,y \in \mathcal{H}_r(\mathbb{K})$ *we have*

$$\mathrm{Tr}((ax + xa^*)y) = \mathrm{Tr}(x(a^*y + ya)).$$

Thus we have $M'_a = M_{a^*}$ *for all a and therefore*

$$\mathfrak{o}(\Lambda_{\mathbb{K}}) = \{M_a : a^* = -a\} \cong \mathfrak{u}(r, \mathbb{K}),$$

$$\mathfrak{p} = \{M_a : a^* = a\}.$$

In the special case $\mathbb{K} = \mathbb{R}$ *(cf. Example 2.14), we have*

$$M_a x = \frac{1}{2}(ax + xa')$$

for all $a \in \mathfrak{gl}(r,\mathbb{R})$ *and* $x \in \mathcal{H}_r(\mathbb{R})$ *(= symmetric real matrices). Further,*

$$\mathfrak{o}(\Lambda_{\mathbb{R}}) = \{M_a : a' = -a\} \cong \mathfrak{o}(r)$$

and

$$\mathfrak{p} = \{M_a : a' = a\}.$$

In all three cases it follows that the Jordan product of $x,y \in \mathcal{H}_r(\mathbb{K})$ *is given by the "anti-commutator"*

$$x \circ y = M_x y = \frac{1}{2}(xy + yx) \tag{2.81}$$

of the ordinary matrix product (recall that multiplication of quaternions is still associative but not commutative). One can show that $S_y = y^{-1}$ *is the inverse in the ordinary matrix sense. In particular, Cramer's rule (for* $\mathbb{K} = \mathbb{R}$ *or* $\mathbb{C}$*) implies that the Jordan algebraic determinant* Δ *is the usual matrix determinant. It follows that* $\Lambda_{\mathbb{K}}$ *has rank r and that the Jordan algebra trace coincides with the usual trace. For* $\mathbb{K} = \mathbb{H}$*, the situation is slightly more complicated (Pfaffian instead of determinant).*

Example 2.82
The forward light cone Λ_n ($n = d + 1 \geq 3$) has the linear automorphism group

$$GL(\Lambda_n) = \mathbb{R}_+ \times SO(1,d)$$

whose Lie algebra

$$g\ell(\Lambda_n) = \mathbb{R} \oplus so(1,d)$$

consists of all (1,d)-blocks of the form

$$M = \begin{pmatrix} \lambda & \mathbf{y}' \\ \mathbf{y} & A + \lambda I \end{pmatrix}$$

with $\mathbf{y} \in \mathbb{R}^d$, $\lambda \in \mathbb{R}$ *and* $A = -A'$. *Thus we have*

$$o(\Lambda_n) = \left\{ \begin{pmatrix} 0 & \mathbf{0} \\ \mathbf{0} & A \end{pmatrix} : A = -A' \right\} \cong o(d)$$

and

$$p = \{M_x : x \in \mathbb{R}^n\}$$

where

$$M_x := \begin{pmatrix} x_0 & \mathbf{x}' \\ \mathbf{x} & x_0 I \end{pmatrix}$$

for all $x = (x_0, \mathbf{x}) \in \mathbb{R}^n$. It follows that the Jordan product is given by

$$x \circ y = (x_0 y_0 + \mathbf{x} \cdot \mathbf{y}, x_0 \mathbf{y} + y_0 \mathbf{x}). \tag{2.83}$$

This product has unit element $e := (1, \mathbf{0})$. *In order to compute the inverse (i.e., the symmetry at e) assume that* $x = (x_0, \mathbf{x})$ satisfies $x_0^2 \neq \mathbf{x} \cdot \mathbf{x}$. The vector

$$x^{-1} := \left(\frac{x_0}{x_0^2 - \mathbf{x} \cdot \mathbf{x}}, \frac{-\mathbf{x}}{x_0^2 - \mathbf{x} \cdot \mathbf{x}} \right) = \frac{1}{x_0^2 - \mathbf{x} \cdot \mathbf{x}} (x_0, -\mathbf{x})$$

clearly satisfies $x \circ x^{-1} = e$ *and is the inverse of x with respect to the Jordan product (2.83). This implies that the Jordan algebra determinant coincides with the quadratic form (2.28). Therefore, the rank is 2 (independent of the dimension n), and we have*

$$\mathrm{tr}(x) = 2x_0$$

[*replace x by* $\lambda e - x$ *in (2.28)*]. *In particular,* $tr(x \circ y) = 2(x_0 y_0 + \mathbf{x} \cdot \mathbf{y})$.

Every Jordan algebra is *power-associative* in the sense that every $x \in \mathbb{R}^n$ generates an associative (and commutative) subalgebra. Thus, the basic results of spectral theory are still available. A nonzero element $c \in \mathbb{R}^n$ is called an *idempotent* if $c^2 := c \circ c = c$. Two idempotents c,d are said to be *orthogonal* if $c \circ d = 0$. Since (2.60) gives

$$\langle c|d\rangle = \langle c \circ c|d\rangle = \langle c|c \circ d\rangle = \langle c|0\rangle = 0,$$

we see that this notion implies geometric orthogonality (but not conversely). An idempotent c is called *minimal* if it is not the sum of two (nonzero) orthogonal idempotents. In this case, $\mathrm{tr}(c^2) = 1$. Every $x \in \mathbb{R}^n$ has a *spectral decomposition* [F1; Theorem III.1.1]

$$x = \sum_{j=1}^{r} \lambda_j c_j, \tag{2.84}$$

where r is the rank, $\lambda_1 \geq \ldots \geq \lambda_r$ are real numbers (uniquely determined), and $c_1, \ldots, c_r$ are minimal orthogonal idempotents satisfying

$$\sum_{j=1}^{r} c_j = e. \tag{2.85}$$

Note that the c_j are not unique unless the λ_j are pairwise distinct. The vector x belongs to Λ if and only if $\lambda_j > 0$ for all j. The spectral resolution (2.84) makes it possible to define *logarithms*

$$\log(x) := \sum_{j=1}^{k} \log(\lambda_j)\, c_j \tag{2.86}$$

and *powers*

$$x^\alpha := \sum_{k=1}^{r} \lambda_j^\alpha c_j$$

as smooth functions of $x \in \Lambda$. One can show that the map $\log : \Lambda \to \mathbb{R}^n$ is a diffeomorphism. Its converse $\exp : \mathbb{R}^n \to \Lambda$ is given by the *exponential mapping*

$$\exp(x) = \sum_{j=1}^{r} e^{\lambda_j} c_j \tag{2.87}$$

PROPOSITION 2.88
For every symmetric cone Λ, the map

$$(s,t) \mapsto (\mathrm{mid}(s,t), s - t) \tag{2.89}$$

is a global diffeomorphism from $\Lambda \times \Lambda$ onto $\Lambda \times \mathbb{R}^n$.

PROOF The formula

$$S_y e = P_y^{1/2} S P_y^{-1/2} e = P_y^{1/2} (y^{-1})^{-1} = y^2$$

implies that $\mathrm{mid}(e,y) = y^{1/2}$ for every $y \in \Lambda$. Since $GL(\Lambda)$ preserves the Riemannian structure on Λ it follows that $s,t \in \Lambda$ have the geodesic middle

$$\mathrm{mid}(s,t) = \mathrm{mid}(P_s^{1/2}e,t) = P_s^{1/2}\,\mathrm{mid}(e,P_s^{-1/2}t) = P_s^{1/2}(P_s^{-1/2}t)^{1/2}.$$

Since $s \mapsto P_s^{\pm 1/2}$ is a smooth map from Λ into $GL(n,\mathbb{R})$, it follows that (2.89) is a C^∞-map. In order to show that it is a diffeomorphism let $(y,\eta) \in \Lambda \times \mathbb{R}^n$ be arbitrary. Since P_y is positive definite, we can form the vector

$$x := P_y^{-1/2}\eta,$$

which depends in a smooth way on η. Its square x^2 belongs to $\bar{\Lambda}$ and $x^2 + 4e$ belongs to Λ since Λ is convex. Using the spectral decomposition of x, it is clear that

$$s_1 := \frac{1}{2}((x^2 + 4e)^{1/2} + x)$$

is the unique vector in Λ satisfying $s_1 - s_1^{-1} = x$. Therefore,

$$s := P_y^{1/2}s_1 = \frac{1}{2}\,[P_y^{1/2}((P_y^{-1/2}\eta)^2 + 4e)^{1/2} + \eta]$$

belongs to Λ and satisfies

$$\eta = P_y^{1/2}x = P_y^{1/2}s_1 - P_y^{1/2}s_1^{-1} = s - P_y^{1/2}SP_y^{-1/2}s = s - S_y s.$$

Also, s is the unique vector in Λ such that $s - S_y s = \eta$. The formula given above for s, and the relation $t = s - \eta$, show that the inverse map $(y,\eta) \mapsto (s,t)$ is C^∞ too. Q.E.D.

Suppose that, for $1 \leq i \leq k$, Λ_i is a symmetric cone in $\mathbb{R}^{n_i}$ of rank r_i. Let Δ_i and tr_i denote the determinant and trace of the Jordan algebra structure on $\mathbb{R}^{n_i}$ associated with Λ_i. The direct product

$$\Lambda := \Lambda_1 \times \cdots \times \Lambda_k$$

is a symmetric cone in $\mathbb{R}^n$, with $n = n_1 + \cdots + n_k$. The corresponding Jordan algebra has the rank

$$r = r_1 + \cdots + r_k,$$

the determinant

$$\Delta(x_1, \ldots, x_k) = \Delta_1(x_1) \cdots \Delta_k(x_k)$$

and the trace

$$tr(x_1, \ldots, x_k) = tr_1(x_1) + \cdots + tr_k(x_k).$$

REMARK 2.90

A symmetric cone Λ is called *irreducible* if Λ is not linearly isomorphic to a product of lower-dimensional symmetric cones. The matrix cones $\Lambda_{\mathbb{K}}$ ($\mathbb{K} = \mathbb{R}, \mathbb{C}, \mathbb{H}$) and the forward light cones Λ_n ($n \geq 3$) are irreducible, and these examples exhaust the list of all irreducible symmetric cones up to one "exceptional" cone of dimension 27 [BK]. Moreover, every symmetric cone is (via an orthogonal change of variables) isomorphic to a direct product of irreducible cones [F1; Proposition III.5.4]. Although some of the arguments in the sequel are slightly easier for irreducible cones, the subsequent analysis is valid for *all* symmetric cones and in particular does not use the (rather deep) classification theory. To avoid confusion over the terminology, let us note that an irreducible symmetric cone Λ is *not* irreducible as a symmetric space [H1] if $\dim \Lambda > 1$. In fact we have $\Lambda = \mathbb{R}_+ \times \Sigma$ where $\mathbb{R}_+$ is the half-line (which is of euclidean type) and $\Sigma := \{y \in \Lambda : \Delta(y) = 1\}$ is an irreducible symmetric space of the noncompact non-euclidean type.

Although the "real" geometry of a symmetric cone $\Lambda \subset \mathbb{R}^n$ is our main concern it is essential to introduce the *right half-space*

$$\Pi := \Lambda + i\mathbb{R}^n \tag{2.91}$$

over Λ, which is a "tube domain" in the complexification $\mathbb{C}^n$ of $\mathbb{R}^n$. Given a vector

$$z = (z_1, \ldots, z_n) \in \mathbb{C}^n,$$

we define

$$z^* := (\bar{z}_1, \ldots, \bar{z}_n). \tag{2.92}$$

This notation is adapted to our basic examples of the matrix cones (Examples 2.14 and 2.21) where (2.92) turns out to be the (hermitian) adjoint of a matrix. In terms of this "involution" we can write

$$\Pi = \{z \in \mathbb{C}^n : z + z^* \in \Lambda\} = \{z \in \mathbb{C}^n : \mathrm{Re}(z) \in \Lambda\}, \tag{2.93}$$

where we put

$$\mathrm{Re}(z) := (\mathrm{Re}(z_1), \ldots, \mathrm{Re}(z_n)) = \frac{z + z^*}{2}. \tag{2.94}$$

Complexifying the Jordan algebra product $x \circ y$ on $\mathbb{R}^n$ induced by Λ, we obtain a complex Jordan algebra on $\mathbb{C}^n$ under the product

$$(x + i\xi) \circ (y + i\eta) := (x \circ y - \xi \circ \eta) + i(\xi \circ y + x \circ \eta) \tag{2.95}$$

defined for all $x, y, \xi, \eta \in \mathbb{R}^n$. The Jordan algebra axioms

$$z \circ w = w \circ z \qquad \text{(commutativity)}$$

and

$$z^2 \circ (z \circ w) = z \circ (z^2 \circ w) \qquad \text{(Jordan identity)}$$

are still valid for all $z, w \in \mathbb{C}^n$, and the unit element $e \in \mathbb{R}^n$ is also the unit element of the complexified algebra. One easily shows that

$$(z \circ w)^* = w^* \circ z^*$$

for all $z, w \in \mathbb{C}^n$.

Example 2.96
Let

$$\mathcal{H}_r(\mathbb{R}) = \{z \in \mathbb{R}^{r\times r} : x' = x\}$$

be the real Jordan algebra of all symmetric real matrices of size $r \times r$, *with anti-commutator product (2.81). Its complexification*

$$\mathcal{H}_r(\mathbb{R})^{\mathbb{C}} = \{z \in \mathbb{C}^{r\times r} : z' = z\} \tag{2.97}$$

is the complex Jordan algebra of all symmetric complex $(r \times r)$*-matrices, with the anti-commutator product. The involution* $z \mapsto z^*$ *is the usual matrix (hermitian) adjoint. The corresponding symmetric cone (2.17) gives rise to the tube domain*

$$\Pi = \{z \in \mathbb{C}^{r\times r} : z' = z, z + z^* > 0\}, \tag{2.98}$$

which is called Siegel's (right) half-space [S4]. *For* $r = 1$, *we obtain the usual (right) half-plane in* $\mathbb{C}$.

Example 2.99
Let

$$\mathcal{H}_r(\mathbb{C}) = \{x \in \mathbb{C}^{r\times r} : x^* = x\}$$

be the real Jordan algebra of all self-adjoint complex matrices of size $r \times r$, *with the anti-commutator product (2.81). Its complexification*

$$\mathcal{H}_r(\mathbb{C})^{\mathbb{C}} = \mathbb{C}^{r\times r} \tag{2.100}$$

is the complex Jordan algebra of all complex $(r \times r)$*-matrices, with the anti-commutator product and the matrix adjoint as involution. The corresponding symmetric cone* $\Lambda = \Lambda_{\mathbb{C}}$ [*cf.* (2.24)] *gives rise to the tube domain*

$$\Pi = \{z \in \mathbb{C}^{r\times r} : z + z^* > 0\}, \tag{2.101}$$

which is called Braun's (right) half-space [B7].

Example 2.102
Let

$$\mathcal{H}_r(\mathbb{H}) = \{x \in \mathbb{H}^{r\times r} : x^* = x\}$$

be the real Jordan algebra of all self-adjoint quaternion matrices of size $r \times r$, *with the anti-commutator product (2.81). Its unit element is the quaternion unit matrix of size r × r. Then one can show that the complexification*

$$\mathcal{H}_r(\mathrm{H})^{\mathrm{C}} \approx \{z \in \mathbb{C}^{2r\times 2r} : z' = z\} \tag{2.103}$$

can be identified with the complex Jordan algebra of all skew-symmetric complex (2r × 2r)-matrices, with unit element

$$e = \begin{pmatrix} 0 & 1 \\ -1 & 0 \end{pmatrix},$$

product

$$z \circ w := \frac{1}{2}(ze^*w + we^*z)$$

and involution $z^* := ez^*e$. *The corresponding cone* $\Lambda = \Lambda_{\mathrm{H}}$ [*cf.* (2.24)] *gives rise to the tube domain*

$$\Pi = \{z \in \mathbb{C}^{2r\times 2r} : z' = -z, ez + z^*e^* > 0\}, \tag{2.104}$$

which is called Krieg's (right) *half-space* [K4].

Every $P \in GL(\Lambda) \subset GL(n,\mathbb{R})$ induces a linear transformation on $\mathbb{C}^n$ via

$$P(x + i\xi) := Px + iP\xi. \tag{2.105}$$

This transformation leaves the subset $\Pi \subset \mathbb{C}^n$ invariant. For fixed $b \in \mathbb{R}^n$, the *translation*

$$T_b(z) := z + ib \tag{2.106}$$

leaves Π invariant. Since $GL(\Lambda)$ is transitive on Λ, it follows that

$$\mathrm{Aff}(\Pi) = \{T_b \circ P : b \in \mathbb{R}^n, P \in GL(\Lambda)\} \tag{2.107}$$

is a transitive group of affine transformations of Π. One can show that every $z \in \Pi$ is invertible in the complex Jordan algebra $\mathbb{C}^n$ and that the inverse z^{-1} still belongs to Π. Thus we obtain a (complex-analytic) *symmetry*

$$S(z) = z^{-1} \tag{2.108}$$

of Π around the base point $e \in \Lambda \subset \Pi$: on $\Lambda \subset \Pi$, it coincides with S as defined in (2.11). One can show that Π has a hermitian structure invariant under the group Aff(Π) that turns it into a hermitian symmetric space and for which S is the geodesic symmetry around e. Then, of course, the symmetry at $y + i\eta \in \Pi$ is given by

$$S_{y+i\eta}(z) = P_y(z - i\eta)^{-1} + i\eta \tag{2.109}$$

for all z. The group Aut(Π) generated by Aff(Π) and S is then the full group of complex automorphisms of Π (cf. [H7]).

3

The Covariance Group of the Fuchs Calculus

The classical Weyl calculus [(0.6) or (1.21)] is covariant under the action of the additive group $\mathbb{R}^{2n}$ on the phase space $\mathbb{R}^{2n}$ (where the symbols live) and its Heisenberg projective representation in $L^2(\mathbb{R}^n)$. Now suppose that Λ is a symmetric cone in $\mathbb{R}^n$, with linear automorphism group

$$GL(\Lambda) := \{P \in GL(n,\mathbb{R}) : P\Lambda = \Lambda\}. \tag{3.1}$$

Let $T^*\Lambda$ denote the *cotangent bundle* over Λ, identified with $\Lambda \times \mathbb{R}^n$ via the standard euclidean structure on $\mathbb{R}^n$. The group $\mathrm{Aut}(T^*\Lambda)$ of all *symplectomorphisms* of $T^*\Lambda$ is an infinite-dimensional Lie group. For every diffeomorphism φ of Λ the codifferential

$$T^*\varphi(y,\eta) = (\varphi(y), d\varphi(y)'^{-1}\eta) \tag{3.2}$$

belongs to $\mathrm{Aut}(T^*\Lambda)$. Here $d\varphi(y) \in GL(n,\mathbb{R})$ is the derivative of φ at $y \in \Lambda$, and $\eta \in \mathbb{R}^n$.

PROPOSITION 3.3
The diffeomorphisms

$$\tilde{P}(y, \eta) := (Py, P'^{-1}\eta) \qquad (P \in GL(\Lambda)), \tag{3.4}$$

$$\tilde{S}(y, \eta) := (Sy, -P_y\eta), \tag{3.5}$$

$$\tilde{T}_b(y, \eta) := (y, \eta + b) \qquad (b \in \mathbb{R}^n), \tag{3.6}$$

$$\tilde{T}^b(y, \eta) := (y, \eta - P_y^{-1}b) \qquad (b \in \mathbb{R}^n) \tag{3.7}$$

of $T^*\Lambda = \Lambda \times \mathbb{R}^n$ *generate a finite dimensional (nonconnected) Lie subgroup* Γ *of* $\mathrm{Aut}(T^*\Lambda)$.

PROOF Each $P \in GL(\Lambda)$ induces a diffeomorphism of Λ with codifferential

$$T^*P(y,\eta) = (Py, P'^{-1}\eta) = \tilde{P}(y,\eta). \tag{3.8}$$

Therefore, $\tilde{P} = T^*P$ belongs to $\mathrm{Aut}(T^*\Lambda)$. Now consider the symmetry $S(y) = y^{-1}$ of Λ. By (2.49), we have

$$(dS)(y) = -P_y^{-1} \tag{3.9}$$

therefore (3.2) implies

$$T^*S(y,\eta) = (y^{-1}, -P'_y\eta)) = (y^{-1}, -P_y\eta) = \tilde{S}(y,\eta).$$

Thus $\tilde{S} = T^*S$ also belongs to $\mathrm{Aut}(T^*\Lambda)$. The same is true of the translations $\tilde{T}_b$ $(b \in \mathbb{R}^n)$. Using (2.44) we obtain

$$\begin{aligned}\tilde{S}\tilde{T}^b\tilde{S}(y,\eta) &= \tilde{S}\tilde{T}_b(y^{-1}, -P'_y\eta)\\ &= \tilde{S}(y^{-1}, b - P_y\eta) = (y, -P_{y^{-1}}(b - P_y\eta)) = (y, \eta - P_y^{-1}b).\end{aligned}$$

Thus,

$$\tilde{T}^b = \tilde{S}\tilde{T}_b\,\tilde{S} \tag{3.10}$$

also belongs to $\mathrm{Aut}(T^*\Lambda)$, and Γ is a subgroup of $\mathrm{Aut}(T^*\Lambda)$. Now let $b,c \in \mathbb{R}^n$ and $P \in GL(\Lambda)$. Define.

$$[P,b,c](y,\eta) := \tilde{T}_b\,\tilde{P}\tilde{T}^c(y,\eta) = (Py,\, b + P'^{-1}(\eta - P_y^{-1}c)). \tag{3.11}$$

Clearly, the "parameters" b,P,c are uniquely determined and depend smoothly on the element $[P,b,c]$ of $\mathrm{Aut}(T^*\Lambda)$.

Applying the formula (2.9) to $Q = P_2$, we see that

$$[P_1,b_1,c_1] \circ [P_2,b_2,c_2] = [P_1P_2,\, b_1 + P'^{-1}_1 b_2,\, P_2^{-1}c_1 + c_2], \tag{3.12}$$

where $\circ$ denotes composition in $\mathrm{Aut}(T^*\Lambda)$. It follows that

$$\Gamma_0 := \{[P,b,c] : b,c \in \mathbb{R}^n,\, P \in GL(\Lambda)\} \tag{3.13}$$

is a closed subgroup of $\mathrm{Aut}(T^*\Lambda)$ and hence a Lie group that is finite dimensional. The "symmetry" $\tilde{S}$ does not belong to (3.13) but satisfies

$$\tilde{S}[P,b,c]\tilde{S} = [P'^{-1}, Pc, P'b], \tag{3.14}$$

since (2.12) and (2.9) imply

$$\begin{aligned}\tilde{S}[P,b,c]\tilde{S}(y,\eta) &= \tilde{S}[P,b,c](y^{-1}, -P_y\eta) \\ &= \tilde{S}(P(y^{-1}), b - P'^{-1}P_y(\eta + c)) \\ &= (P'^{-1}y, Pc + P(\eta - P_y^{-1}P'b)).\end{aligned} \tag{3.15}$$

It follows that

$$\Gamma := \Gamma_0 \cup \tilde{S}\,\Gamma_0 \qquad \text{(disjoint union)} \tag{3.16}$$

is a (nonconnected) Lie group. Q.E.D.

PROPOSITION 3.17
The group Γ *defined in Proposition 3.3 has a continuous unitary representation* U_∞ *into* $L^2(\Lambda)$ *such that for* $u \in L^2(\Lambda)$ *and* $t \in \Lambda$,

$$U_\infty([P, b, c])u(t) = u(P^{-1}t)e^{2\pi i(\langle b,t\rangle + \langle c,St\rangle)} \tag{3.18}$$

for all $b,c \in \mathbb{R}^n$ *and* $P \in GL(\Lambda)$, *and*

$$U_\infty(\tilde{S})u(t) = u(t^{-1}). \tag{3.19}$$

PROOF Use the fact that dm is invariant under $GL(\Lambda)$ and S, as well as (3.14). Q.E.D.

DEFINITION 3.20 Let

$$G := GL(\Lambda) \times \mathbb{R}^n \tag{3.21}$$

be the semi-direct product of $GL(\Lambda)$ by $\mathbb{R}^n$, consisting of all pairs $\gamma = (P,b)$ with $P \in GL(\Lambda)$, $b \in \mathbb{R}^n$ and having the composition

$$(P_1,b_1)(P_2,b_2) = (P_1P_2\,, b_1 + P_1'^{-1}b_2). \tag{3.22}$$

Identifying G with a subgroup of Γ_0 by means of the map $(P,b) \mapsto (P,b,0)$, we let G act on $T^*\Lambda = \Lambda \times \mathbb{R}^n$ by

$$[P,b](y,\eta) := (Py\,, P'^{-1}\eta + b). \tag{3.23}$$

Also, let us denote as U the restriction of the representation U_∞ to G. Then

$$U(P,b)u(t) = u(P^{-1}t)\, e^{2i\pi\langle b,t\rangle} \tag{3.24}$$

for all $(P,b) \in G$.

PROPOSITION 3.25

The Fuchs calculus is covariant under the representation U_∞ *of the group* $\Gamma \subset \mathrm{Aut}(T^*\Lambda)$ *on* $L^2(\Lambda)$ *and its action (3.11) on the phase space* $T^*\Lambda = \Lambda \times \mathbb{R}^n$. *Thus, for all* $\gamma \in \Gamma$, *the operator with active (resp., passive) symbol* $f \circ \gamma^{-1}$ *is* $U_\infty(\gamma)\, A U_\infty(\gamma^{-1})$ *if* f *is the active (resp., passive) symbol of the operator* A.

In particular, for $P \in GL(\Lambda)$ *and* $b \in \mathbb{R}^n$, *it follows that* $f \circ [P,b]^{-1}$ *corresponds to* $U(P,b)\, A\, U((P,b)^{-1})$ *for both types of symbols.*

PROOF As a consequence of (3.24) and (1.20), and with

$$v(t) = u(Pt)\, e^{-2i\pi\langle b,Pt\rangle},$$

one may write

$$\begin{aligned}[U(P,b)\sigma_{y,\eta}\, U(P,b)^{-1}u](t) &= e^{2i\pi\langle b,t\rangle}\, v(2y - P^{-1}t)\, e^{4i\pi\langle \eta, P^{-1}t - y\rangle} \\ &= u(2Py - t)\, e^{4i\pi\langle b+P'^{-1}\eta,\ t-Py\rangle},\end{aligned}$$

i.e.,

$$U(P,b)\sigma_{y,\eta}U(P,b)^{-1} = \sigma_{Py,\ b+P'^{-1}\eta} = \sigma_{[P,b](y,\eta)}. \tag{3.26}$$

In just the same way, one sees that

$$\sigma_{e,0}\sigma_{y,\eta}\sigma_{e,0} = \sigma_{\tilde{S}(y,\eta)},$$

so that Proposition 3.25 is proved whenever $\gamma = \tilde{S}$ or γ is of the type (P,b), in view of the basic Definitions 1.6 and 1.10 of the operators with given active or passive Fuchs symbol. It is proved in general as a consequence of (3.10) and of the definition of Γ given in Proposition 3.3.

Q.E.D.

Since the assumption of Proposition 1.14 is satisfied (by Proposition 2.88) we can describe, at least formally, the link between the active and passive symbols of an operator. Define the positive function ω on Λ by

$$dm(t) = \omega(t)^{-1}\, dt, \tag{3.27}$$

where dt is the standard Lebesgue measure on $\mathbb{R}^n$ and dm(t) is the $GL(\Lambda)$-

invariant measure on Λ already considered, renormalized if necessary so that $\omega(e) = 1$. Then $\omega(t) = \mathrm{Det}\, P_t^{1/2}$.

PROPOSITION 3.28
For $f \in C_0^\infty(T^*\Lambda)$, $\mathrm{Op}(f)$ *is of trace-class in* $L^2(\Lambda)$, *and the passive symbol* g *is related to* f *by a convolution operator in the fibers of* $T^*\Lambda$. *Namely, one has*

$$g(y, \eta) = F\left(\frac{1}{2i\pi} P_y^{-1/2} \frac{\partial}{\partial \eta}\right) f(y, \eta), \tag{3.29}$$

where F *is the (radial) function on* $\mathbb{R}^n$ *uniquely determined by*

$$F(St - t) = 2^{2n}\omega(t)^{-2}\left[\mathrm{Det}\left(\frac{\partial}{\partial x}(S_x t)\right)(x = e) \quad \mathrm{Det}\left(\frac{\partial}{\partial t}(t - St)\right)\right]^{-1} \tag{3.30}$$

for all $t \in \Lambda$.

PROOF We omit the proof of the first point here, since a considerably more precise result will be proved later in an independent way; namely, after we have introduced in Definition 8.13 the "Schwartz spaces" $\mathcal{S}(\Lambda)$ and $\mathcal{S}'(\Lambda)$, we shall be able to state that, under the present assumption on f, $\mathrm{Op}(f)$ extends as a continuous operator from $\mathcal{S}'(\Lambda)$ to $\mathcal{S}(\Lambda)$. The proof of this latter fact follows from Proposition 12.10. Since $dt/dm(t) = \omega(t)$, the kernel of $\mathrm{Op}(f)$ with respect to the measure $dm(t)$ is $\omega(t)\, k(s,t)$, where $k(s,t)$ has been defined in (1.16). Choose an orthonormal basis (u_j) of $L^2(\Lambda)$ so that, according to Definition 1.10, one has

$$\begin{aligned} g(y, \eta) &= 2^n \sum_j (u_j | \mathrm{Op}(f)\sigma_{y,\eta} u_j) \\ &= 2^n \sum_j \int \bar{u}_j(s)\, dm(s) \int k(s, t) u_j(S_y t) e^{2i\pi\langle \eta, t - S_y t\rangle} dt. \end{aligned}$$

Since

$$k(S_y t, t) = \sum_j u_j(S_y t) \int k(s, t) \bar{u}_j(s)\, dm(s),$$

one thus has

$$g(y, \eta) = 2^n \int_\Lambda k(S_y t, t) e^{2i\pi\langle \eta, t - S_y t\rangle} \omega(t)\, dm(t). \tag{3.31}$$

Note that

$$\mathrm{Det}\left(\frac{\partial}{\partial t}\,\mathrm{mid}(s, t)\right) = \left(\mathrm{Det}\left(\frac{\partial}{\partial x}\,(S_x s.)\right)(x = \mathrm{mid}(s, t))\right)^{-1} \tag{3.32}$$

so that (1.16) yields

$$k(St, t) = 2^n(\mathcal{F}_2^{-1}f)(e, St - t)\left(\mathrm{Det}\left(\frac{\partial}{\partial x}\,(S_x St)(x = e)\right)\right)^{-1}. \tag{3.33}$$

Obviously,

$$\omega(Pt) = (\det P)\,\omega(t) = \omega(Pe)\,\omega(t) \tag{3.34}$$

if $P \in GL(\Lambda)$. By (3.9) the derivative of the map $t \mapsto St$ is the map $h \mapsto -P_t^{-1}h$. Since S is dm-preserving, one has

$$\omega(St) = \omega(t)\,\left|\mathrm{Det}\,\frac{\partial}{\partial t}\,(St)\right| = \omega(t)\,\mathrm{Det}\,P_t^{-1} = \omega(t)^{-1} \tag{3.35}$$

and

$$\omega(S_y t) = \omega(P_y St) = \omega(y)^2\,\omega(t)^{-1}. \tag{3.36}$$

Given $s,x \in \Lambda$, one has $S_{S_x s}Ss = SS_x s$, so that if B is a small geodesic ball centered at e, the dm-measures of the sets $\{S_x s : x \in B\}$ and $\{S_y Ss : y \in B\}$ coincide for fixed s. Therefore,

$$\omega(s)\mathrm{Det}\left(\frac{\partial}{\partial x}\,(S_x s)\right)(x = e) = \omega(s)^{-1}\mathrm{Det}\left(\frac{\partial}{\partial x}\,(S_x Ss)\right)(x = e). \tag{3.37}$$

Evaluating the jacobian on the right-hand side of (3.33) and letting $s = St$, we obtain

$$\left(\mathrm{Det}\left(\frac{\partial}{\partial x}\,(S_x s)\right)(x = e)\right)^{-1} = \omega(t)^{-2}\left(\mathrm{Det}\left(\frac{\partial}{\partial x}\,(S_x t)\right)(x = e)\right)^{-1}. \tag{3.38}$$

From (3.31) and (3.33), we obtain

$$g(e, \eta) = 2^{2n}\int (\mathcal{F}_2^{-1}f)(e, St - t)\omega(t)^{-2} \cdot \left(\mathrm{Det}\left(\frac{\partial}{\partial x}\,(S_x t)\right)(x = e)\right)^{-1} e^{2i\pi\langle\eta, t - St\rangle}dt. \tag{3.39}$$

Now a consequence of Proposition 2.88 is that the map $t \mapsto St - t$ from Λ to $\mathbb{R}^n$ is a diffeomorphism onto. Thus a change of variable in (3.39) gives the asserted result when $y = e$, and covariance under $GL(\Lambda)$ yields it in general, as it implies that the map $f \mapsto g$ commutes with the transformation $(y,\eta) \mapsto (Py, P'^{-1}\eta)$. Q.E.D.

For irreducible cones $\Lambda \subset \mathbb{R}^n$ of rank r, the density function ω and the Jordan algebra determinant Δ are related by

$$dm(t) = \Delta(t)^{-n/r}\, dt. \tag{3.40}$$

This follows from the fact that

$$\Delta(Px) = (\mathrm{Det}\, P)^{r/n} \cdot \Delta(x) \tag{3.41}$$

for all $P \in GL(\Lambda)$ [BK]. In the reducible case $\Lambda = \prod_i \Lambda_i$ we may, with obvious notations, take

$$dm(t) = \prod_i \Delta_i(t_i)^{-n_i/r_i}\, dt_i. \tag{3.42}$$

4

Geometric Inequalities

The development of pseudodifferential analysis on a symmetric cone Λ is based on estimates and on a careful accounting of how uniform these are; they may depend on Λ, on points of Λ or Π, on symbols, on weight functions, on extra parameters, etc. Constants $C > 0$ enter the inequalities and so do some exponents N. They may vary from one formula to the next one. However, with the exception that all constants are assumed to depend on Λ (in particular, on its dimension and rank), we shall always display explicitly all the additional data on which they depend. Accordingly, constants such as C depend only on Λ. We also set

$$f(s) \lesssim g(s) \ (s \in S) \tag{4.1}$$

if we have

$$f(s) \leq C\, g(s) \tag{4.2}$$

for all $s \in S$. We write

$$f(s) \sim g(s) \ (s \in S) \tag{4.3}$$

if C may be chosen such that

$$C^{-1} g(s) \leq f(s) \leq C\, g(s) \tag{4.4}$$

for all $s \in \Lambda$.

DEFINITION 4.5 Let Λ be a symmetric cone in $\mathbb{R}^n$ of rank r. Let Δ be the determinant function of Λ. Define a smooth function $\delta: \Lambda \times \Lambda \to \mathbb{R}$ by putting

$$\delta(s,t) := 2^{-r}\frac{\Delta(s + t)}{\Delta(s)^{1/2}\,\Delta(t)^{1/2}} \tag{4.6}$$

for all $s,t \in \Lambda$.

Since the determinant function is multiplicative with respect to direct products [cf. (2.74)] we have

$$\delta(s,t) = \prod_i \delta_i(s_i,t_i) \tag{4.7}$$

in the reducible case.

LEMMA 4.8
For all $P \in GL(\Lambda)$ *and* $s,t \in \Lambda$, *we have*

$$\delta(Ps,Pt) = \delta(s,t).$$

PROOF By (4.7) we may assume that Λ is irreducible. Then the assertion follows from (3.41). Q.E.D.

LEMMA 4.9
For all $s,t \in \Lambda$ *we have* $\delta(s,t) \geq 1$.

PROOF By Lemma 4.8 we may assume $t = e$. Let

$$s = \sum_{j=1}^{r} \lambda_j c_j \tag{4.10}$$

be the spectral decomposition of s. Since $\Sigma c_j = e$, we obtain

$$\delta(s,e) = 2^{-r}\prod_j \frac{\lambda_j + 1}{\lambda_j^{1/2}} \geq 1. \tag{4.11}$$

Q.E.D.

LEMMA 4.12
For all $s,t,y \in \Lambda$ *we have*

$$\delta(s,t) \leq 2^r\,\delta(s,y)\,\delta(y,t)$$

where r *is the rank of* Λ.

PROOF By invariance under $GL(\Lambda)$ we may assume that $y = e$. In this case we have to show

$$\Delta(s + t) \leq \Delta(e + s)\,\Delta(e + t). \tag{4.13}$$

We prove the stronger inequality (for all $\theta \geq 0$)

$$\log \Delta(e + s + \theta t) \leq \log \Delta(e + s) + \log \Delta(e + \theta t). \tag{4.14}$$

Since (4.14) holds for $\theta = 0$, it suffices to prove

$$\frac{\partial}{\partial\theta} \log \Delta(e + s + \theta t) \leq \frac{\partial}{\partial\theta} \log \Delta(e + \theta t).$$

Applying [K2; Lemma 2.3.4], we have to show that the Jordan product satisfies

$$\mathrm{tr}((e + s + \theta t)^{-1} \circ t) \leq \mathrm{tr}((e + \theta t)^{-1} \circ t). \tag{4.15}$$

Since $t \in \Lambda$, (4.15) follows from the fact that

$$(e + \theta t)^{-1} - (e + s + \theta t)^{-1} \in \bar{\Lambda} \qquad \text{(closure of } \Lambda\text{)},$$

which is seen as follows: Putting

$$x \leq y \text{ iff } y - x \in \bar{\Lambda}, \tag{4.16}$$

we have

$$x \leq y \Rightarrow y^{-1} \leq x^{-1} \tag{4.17}$$

for all $x,y \in \Lambda$. For $x = e$, this follows from the spectral decomposition of y. In general, use

$$x \leq y \Rightarrow e \leq P_x^{-1/2} y \Rightarrow (P_x^{-1/2} y)^{-1} = P_x^{1/2} y^{-1} \leq e \Rightarrow y^{-1} \leq P_x^{-1/2} e = x^{-1}.$$

Q.E.D.

LEMMA 4.18
For $s,y \in \Lambda$ *we have*

$$\delta(s,y) \sim \delta(s,S_y s)^{1/2}.$$

PROOF Since $S_y = P_y S P_y^{-1}$ by (2.13), and δ is $GL(\Lambda)$-invariant by Lemma 4.8, we may assume $y = e$. In terms of the spectral decomposition (4.10) of s, we obtain

$$Ss = s^{-1} = \sum_j \lambda_j^{-1} c_j. \tag{4.19}$$

Therefore

$$\delta(s, s^{-1}) = 2^{-r} \prod_j \frac{\lambda_j + \lambda_j^{-1}}{\lambda_j^{1/2}\lambda_j^{-1/2}} = \prod_j \frac{\lambda_j + \lambda_j^{-1}}{2}.$$

Since $\lambda + \lambda^{-1} \leq (\lambda^{1/2} + \lambda^{-1/2})^2 \leq 2(\lambda + \lambda^{-1})$, a comparison with (4.11) yields the assertion. Q.E.D.

Given a symmetric cone $\Lambda \subset \mathbb{R}^n$, the manifold $\Lambda \times \mathbb{R}^n$ can be identified both with the tangent bundle $T\Lambda$ and the cotangent bundle $T^*\Lambda$. At the base point $e \in \Lambda$, we may even identify

$$T_e\Lambda = \mathbb{R}^n = T_e^*\Lambda$$

as euclidean spaces, endowed with the standard euclidean norm $|\cdot|$ on $\mathbb{R}^n$. However, the group $GL(\Lambda)$ acts on $T\Lambda$ and $T^*\Lambda$ in a different way. For $P \in GL(\Lambda)$ we have for all $(y,\eta) \in \Lambda \times \mathbb{R}^n$

$$(TP)(y,\eta) = (Py,P\eta) \tag{4.20}$$

whereas

$$(T^*P)(y,\eta) = (Py,P'^{-1}\eta). \tag{4.21}$$

Accordingly we define two "fiber norms" on $\mathbb{R}^n$ by putting

$$\|x\|_{Pe} := |P^{-1}x|, \tag{4.22}$$

$$|\xi|_{Pe} := |P'\xi| \tag{4.23}$$

for all $x,\xi \in \mathbb{R}^n$. Here $P \in GL(\Lambda)$ and the numbers (4.22), (4.23) depend only on $Pe \in \Lambda$ since the stabilizer subgroup $O(\Lambda)$ at e preserves the euclidean norm $|\cdot|$. Using the "quadratic representation" operators (2.8) and (2.66) we may write

$$\|x\|_y = |P_y^{-1/2}x|, \tag{4.24}$$

$$|\xi|_y = |P_y^{1/2}\xi| \tag{4.25}$$

for all $y \in \Lambda$. This shows that (4.24) and (4.25) define smooth fiber norms on $T\Lambda$ and $T^*\Lambda$, respectively, which are invariant under the natural actions of $GL(\Lambda)$. Moreover, for every $y \in \Lambda$, (4.25) is the dual norm of (4.24) in the sense that

$$|\xi|_y = \sup_{\|x\|_y \leq 1} \langle x,\xi\rangle. \tag{4.26}$$

LEMMA 4.27
For $s,y \in \Lambda$ *we have (in the irreducible* rank r *case)*

$$(1 + \|s - S_y s\|_y^2)^{1/2} \lesssim \delta(s,S_y s) \lesssim (1 + \|s - S_y s\|_y^2)^{r/2}. \tag{4.28}$$

PROOF By $GL(\Lambda)$-invariance of (4.24) and δ, we may assume $y = e$. Using the spectral decomposition (4.10) and (4.19) of s and s^{-1}, we obtain

$$\begin{aligned}
1 + \frac{1}{4}\operatorname{tr}((s - s^{-1})^2) &= 1 + \sum_j \left(\frac{\lambda_j - \lambda_j^{-1}}{2}\right)^2 \\
&\leq \prod_j \left[1 + \left(\frac{\lambda_j - \lambda_j^{-1}}{2}\right)^2\right] \\
&= \prod_j \left(\frac{\lambda_j + \lambda_j^{-1}}{2}\right)^2 = \delta(s, s^{-1})^2 \\
&\leq \left[1 + \sum_j \left(\frac{\lambda_j - \lambda_j^{-1}}{2}\right)^2\right]^r = \left[1 + \frac{1}{4}\operatorname{tr}(s - s^{-1})^2\right]^r.
\end{aligned}$$

Lemma 4.27 thus follows the equivalence of $|x|$ with the norm defined in (2.78). Q.E.D.

LEMMA 4.29
For all $s,t \in \Lambda$ *and* $\xi \in \mathbb{R}^n$ *we have*

$$\|\xi\|_t \lesssim \delta(s,t)^2\, \|\xi\|_s. \tag{4.30}$$

PROOF We may assume $t = e$ by $GL(\Lambda)$-invariance. Using (4.25) we have to show that

$$|P_s^{1/2}\,\xi| \lesssim \delta(s,e)^2\,|\xi| \tag{4.31}$$

for all $s \in \Lambda$ and $\xi \in \mathbb{R}^n$, i.e., that the operator $P_s^{1/2}$ has norm $\|P_s^{1/2}\| \lesssim \delta(s,e)^2$. Since $s \mapsto P_s$ is a quadratic polynomial by (2.66) we have (using the spectral decomposition of s)

$$\|P_s\| \lesssim |s|^2 \sim \sum_j \lambda_j^2 \lesssim \prod_j (\lambda_j^{1/2} + \lambda_j^{-1/2})^4 \lesssim \delta(s, e)^4. \tag{4.32}$$

Since P_s is self-adjoint, we get $\|P_s^{1/2}\| = \|P_s\|^{1/2} \lesssim \delta(s,e)^2$.

Q.E.D.

LEMMA 4.33
For $s,t \in \Lambda$ *we have*

$$\Delta(s) \lesssim \Delta(t) \cdot \delta(s,t)^2. \tag{4.34}$$

PROOF Since $\Delta(s)/\Delta(t)$ is $GL(\Lambda)$-invariant we may assume that $t = e$. Now the spectral decomposition (4.10) of s gives

$$\delta(s, e) = \prod_j \frac{\lambda_j^{1/2} + \lambda_j^{-1/2}}{2} \geq 2^{-r} \prod_j \lambda_j^{1/2} = 2^{-r}\Delta(s)^{1/2}.$$

Q.E.D.

LEMMA 4.35
For $s \in \Lambda$ *and* $h \in \mathbb{R}^n$, *we have*

$$|s \circ h| \gtrsim \delta(e,s)^{-2}\,|h|, \tag{4.36}$$

i.e., the Jordan multiplication operator M_s *satisfies*

$$\|M_s^{-1}\| \lesssim \delta(e,s)^2. \tag{4.37}$$

PROOF For every $x \in \bar{\Lambda}$ the multiplication operator M_x is positive semidefinite [BK]. It follows from (2.66) that

$$\begin{aligned} 2|s \circ h|^2 &= 2\langle M_s h, M_s h\rangle \\ &= 2\langle M_s^2 h, h\rangle \\ &= \langle P_s h, h\rangle + \langle M_{s^2} h, h\rangle \end{aligned}$$

$$\begin{aligned} &\geq \langle P_s h,h\rangle \\ &= \langle P_s^{1/2}h, P_s^{1/2}h\rangle \\ &\geq \|P_s^{-1/2}\|^{-2} \cdot \langle h,h\rangle. \end{aligned}$$

Applying (4.31) to s^{-1}, we obtain

$$\|P_s^{1/2}\| \lesssim \delta(e,s^{-1})^2 = \delta(e,s)^2, \tag{4.38}$$

and, therefore,

$$|s \circ h| \gtrsim \delta(e,s)^{-2} \cdot |h|.$$

Q.E.D.

LEMMA 4.39
Let d *be the Riemannian distance on* Λ. *Then for every* $s,t \in \Lambda$ *we have*

$$1 + d(s,t) \sim 1 + \log \delta(s,t),$$

which amounts, since $\delta(s,t) \geq 1$, *to*

$$C^{-1}\,\delta(s,t)^{1/N} \leq e^{d(s,t)} \leq C\,\delta(s,t)^N \tag{4.40}$$

with a pair (C,N) *depending only on* Λ.

PROOF Since δ and d are both invariant under $GL(\Lambda)$, acting on $\Lambda \times \Lambda$, we may assume $t = e$. Let $s \in \Lambda$ have the spectral decomposition (4.10). Define

$$x := \sum_{j=1}^{r} \ln(\lambda_j) \cdot c_j.$$

Since M_x belongs to the subspace $\wp$ of the Cartan decomposition of the Lie algebra $\mathfrak{gl}(\Lambda)$ of $GL(\Lambda)$, it follows that

$$\varphi(\theta) := \exp(\theta \cdot M_x)(e)$$

is a geodesic in Λ. The minimal projections $c_1, \ldots, c_r$ generate a commutative and associative unital subalgebra $A \approx \mathbb{R}^r$ (direct product) of $\mathbb{R}^n$, which contains x. Computing in A, we obtain

$$\varphi(\theta) = \sum_{j=1}^{r} \lambda_j^{\theta} c_j$$

and

$$P_{\varphi(\theta)}^{-1/2}\dot{\varphi}(\theta) = \varphi(\theta)^{-1/2}\dot{\varphi}(\theta)\varphi(\theta)^{-1/2}$$

$$= \sum_{j=1}^{r} \lambda_j^{-\theta/2} \ln(\lambda_j)\lambda_j^{\theta}\lambda_j^{-\theta/2}c_j = \sum_{j=1}^{r} \ln(\lambda_j)c_j = x.$$

It follows that

$$d(s, e) = \int_0^1 \|\dot{\varphi}(\theta)\|_{\varphi(\theta)} d\theta$$

$$= \int_0^1 |P_{\varphi(\theta)}^{-1/2}\dot{\varphi}(\theta)|\, d\theta = |x| \sim \sum_{j=1}^{r} |\ln \lambda_j|. \tag{4.41}$$

Since

$$\delta(s, e)^2 \sim \prod_{j=1}^{r} \max\left(\lambda_j, \frac{1}{\lambda_j}\right) = \exp\left(\sum_{j=1}^{r} |\ln \lambda_j|\right),$$

the assertion follows. Q.E.D.

The Jordan algebra determinant $\Delta : \mathbb{R}^n \to \mathbb{R}$ has a complex-analytic extension

$$\Delta : \mathbb{C}^n \to \mathbb{C},$$

which is a polynomial in $z_1, \ldots, z_n$ of degree r. Here r is the rank of Λ. For a direct product $\Lambda = \prod \Lambda_i$, $\Pi = \prod \Pi_i$ is the corresponding half-space and the complex determinant function is still multiplicative:

$$\Delta(z) = \prod_i \Delta_i(z_i) \tag{4.42}$$

for all $z \in \mathbb{C}^n$. Define a function $\delta : \Pi \times \Pi \to \mathbb{R}_+$ by putting

$$\delta(z, w) := \frac{|\Delta(z + w^*)|}{\Delta(z + z^*)^{1/2}\Delta(w + w^*)^{1/2}}$$

$$= 2^{-r} \frac{|\Delta(z + w^*)|}{\Delta(\mathrm{Re}(z))\Delta(\mathrm{Re}(w))} \tag{4.43}$$

for all $z,w \in \Pi$. Then we have

$$\delta(z, w) = \prod_i \delta_i(z_i, w_i) \tag{4.44}$$

in the reducible case.

LEMMA 4.45
For all $\gamma \in \mathrm{Aff}(\Pi)$ *and* $z,w \in \Pi$, *we have*

$$\delta(\gamma z,\gamma w) = \delta(z,w).$$

PROOF By (2.107), the group $\mathrm{Aff}(\Pi)$ is generated by the translations T_b ($b \in \mathbb{R}^n$) and the (complexified) transformations $P \in GL(\Lambda)$. For $\gamma = T_b$, the assertion is trivial. For $\gamma = P$, apply (3.41). Q.E.D.

LEMMA 4.46
For all $z,w \in \Pi$, *we have*

$$\delta(z,w) \geq \delta(\mathrm{Re}(z),\mathrm{Re}(w)). \tag{4.47}$$

PROOF Write $z = x + i\xi$, and $w = y + i\eta$. Then $x \in \Lambda$, and (3.41) implies

$$\Delta(z) = \Delta(P_x^{1/2}(e + iP_x^{-1/2}\xi)) = \Delta(x)\,\Delta(e + iP_x^{-1/2}\xi).$$

Substituting $z + w^*$ for z, we obtain

$$\delta(z,w) = \delta(x,y)\,|\Delta(e + iP_{x+y}^{-1/2}(\xi - \eta))|. \tag{4.48}$$

If $s \in \mathbb{R}^n$ has the spectral resolution (4.10), then

$$|\Delta(e + is)| = \prod_{j=1}^{r} |1 + i\lambda_j| \geq 1. \tag{4.49}$$

Therefore (4.48) yields the assertion. Q.E.D.

As a consequence of Lemma 4.9, one has $\delta \geq 1$ on $\Pi \times \Pi$.

LEMMA 4.50
For $z = x + i\xi$, $w = y + i\eta \in \Pi$, *we have*

$$1 + \log(\delta(x,y) + \|\xi - \eta\|_x^2) \sim 1 + \log \delta(z,w) \sim 1 + d(z,w) \tag{4.51}$$

where $d(z,w)$ *is the geodesic distance in* Π.

PROOF Let

$$s := P_{x+y}^{-1/2}(\xi - \eta)$$

have the spectral resolution (4.10). Then

$$1 + \mathrm{tr}(s^2) \lesssim |\Delta(e + is)|^2 \lesssim [1 + \mathrm{tr}(s^2)]^r$$

since

$$1 + \sum_{j=1}^{r} \lambda_j^2 \leq \prod_{j=1}^{r} (1 + \lambda_j^2) \leq \left[1 + \sum_{j=1}^{r} \lambda_j^2\right]^r.$$

By (4.24) this implies

$$1 + \|\xi - \eta\|_{x+y}^2 \lesssim |\Delta(e + is)|^2 \lesssim (1 + \|\xi - \eta\|_{x+y}^2)^r. \tag{4.52}$$

For all $x,y \in \Lambda$ we have

$$\delta(x,x + y) \lesssim \delta(x,y) \tag{4.53}$$

as a reduction to the case $x = e$ and a spectral decomposition show. With Lemma 4.29, we obtain

$$\begin{aligned}\delta(x, y)^{-2}\|\xi - \eta\|_x &\lesssim \delta(x, x + y)^{-2}\|\xi - \eta\|_x \lesssim \|\xi - \eta\|_{x+y} \\ &\lesssim \delta(x, x + y)^2\|\xi - \eta\|_x \lesssim \delta(x, y)^2\|\xi - \eta\|_x.\end{aligned} \tag{4.54}$$

Now combine (4.48), (4.52), and (4.54) to obtain

$$\begin{aligned}1 + \log \delta(z, w) &= 1 + \log \delta(x, y) + \log|\Delta(e + is)| \\ &\sim 1 + \log \delta(x, y) + \log(1 + \|\xi - \eta\|_{x+y}^2) \\ &\sim 1 + \log \delta(x, y) + \log(1 + \|\xi - \eta\|_x^2) \\ &\sim 1 + \log(\delta(x, y) + \|\xi - \eta\|_x^2)\end{aligned}$$

since $\delta(x,y) \geq 1$. This proves the first equivalence in (4.51).

In order to prove the second equivalence, consider the holomorphic automorphism group Aut(Π) of Π. It is known [K2, FK2, L2] that the Bergman kernel of Π is given by

$$E(z,w) = \Delta(z + w^*)^{-p} \tag{4.55}$$

for all z, w $\in \Pi$, where p is a numerical constant called the *genus* (cf. Section 14). We have the transformation formula

$$\text{Det}\, \partial\gamma(z)\, E(\gamma(z), \gamma(w))\, \overline{\text{Det}\, \partial\gamma(w)} = E(z, w) \tag{4.56}$$

for all $\gamma \in \text{Aut}(\Pi)$. Here $\partial\gamma(z) \in GL(n,\mathbb{C})$ is the complex derivative. It follows that

$$\delta(z, w) = \left| \frac{E(z, w)}{E(z, z)^{1/2} E(w, w)^{1/2}} \right|^{-1/p} \tag{4.57}$$

satisfies

$$\delta(\gamma(z),\gamma(w)) = \delta(z,w) \tag{4.58}$$

for all $\gamma \in \text{Aut}(\Pi)$. Since d(z,w) is also Aut(Π)-invariant, it suffices to show that

$$1 + \log \delta(z,e) \sim 1 + d(z,e) \tag{4.59}$$

for all $z \in \Pi$. The stabilizer group

$$K_\Pi := \{\gamma \in \text{Aut}(\Pi) : \gamma(e) = e\} \tag{4.60}$$

satisfies

$$\Pi = \{\gamma(x) : \gamma \in K_\Pi, x \in \Lambda\}$$

as follows easily by considering the bounded realization of Π (cf. Section 16). Therefore, we may assume $z \in \Lambda$. Since Λ is a geodesically embedded submanifold of Π, its geodesic distance agrees with the restriction of d, and the assertion follows from Lemma 4.39. Q.E.D.

LEMMA 4.61
There exists a constant $c_0 > 0$ *such that for* $z = x + i\xi \in \Pi$, $w = y + i\eta \in \Pi$ *we have*

$$|z - w| \leq \frac{1}{c_0}\delta(e,x)^{-2} \Rightarrow \delta(z,w) \leq c_0. \tag{4.62}$$

PROOF Applying (4.38) we obtain

$$|e - P_x^{-1/2}y| = |P_x^{-1/2}(x - y)| \leq \|P_x^{-1/2}\| \cdot |x - y| \lesssim \delta(e,x)^2 \cdot |z - w|.$$

It follows that we may choose c_0 such that

$$\delta(e,x)^2 \cdot |z - w| \leq \frac{1}{c_0} \Rightarrow (\mathrm{tr}(e - P_x^{-1/2}y))^{1/2} \leq \frac{1}{2}.$$

This implies that

$$\delta(x,y) = \delta(e,P_x^{-1/2}y)$$

is bounded. Similarly,

$$\|\xi - \eta\|_x \lesssim \delta(e,x)^2 \cdot |\xi - \eta| \leq \delta(e,x)^2 \cdot |z - w| \leq \frac{1}{c_0}$$

is bounded. Now apply (4.51). Q.E.D.

REMARK 4.63

We have seen that Definition 1.6 (of the Fuchs calculus) applies as well to $\mathbb{R}^n$ (in which case it is the Weyl calculus) as to symmetric cones Λ, or to products $\Lambda \times \mathbb{R}^p$, since these are also symmetric domains: only the group $GL(\Lambda)$ is replaced by a group of affine (not linear) transformations of $\mathbb{R}^{n+p}$, a fact that explains why one only gets a projective (i.e., up to constant factors of modulus 1) representation, because of the Heisenberg component. We claim that all the results of this paper are valid for such products except (cf. Theorem 0.27) that, in the main estimates, powers of $e^{d(z,w)}$ have to be replaced by powers of $1 + |z - w|^2$ for the $\mathbb{R}^p$ factor. To avoid very cumbersome notations, we shall state and prove results only in the case of symmetric cones. The case of the half-space $\mathbb{R}_+^{n+1}$, however, has been made explicit in [U6; Section 16], and some of its features will be recalled shortly at the end of Section 13.

The following two Lemmas will be needed in Section 12. It is convenient here to assume that one has substituted (cf. Section 2) the (equivalent) norm $x \mapsto (\mathrm{tr}(x^2))^{1/2}$ for the original $x \mapsto |x|$.

LEMMA 4.64

For every $P \in GL(\Lambda)$ *there exists* $b \in \partial\Lambda$ *(the boundary of* Λ*) of norm* 1 *such that* $|P'b| = \|P^{-1}\|^{-1}$.

PROOF Since $P = P_x Q$ where $x \in \Lambda$ and $Q \in O(\Lambda)$ is an isometry, we may assume $P = P_x$. In terms of a spectral decomposition

$$x = \sum_{i=1}^{r} \lambda_i c_i, \qquad \lambda_1 \geq \cdots \geq \lambda_r > 0$$

we have $\|P_x^{-1}\| = \lambda_r^{-2}$ and, for $b := c_r$, $P_x'c_r = P_x c_r = \lambda_r^2 c_r$. Thus $|P_x'c_r| = \lambda_r^2 = \|P_x^{-1}\|^{-1}$. Q.E.D.

LEMMA 4.65
For every $b \in \mathbb{R}^n \backslash \{0\}$ *there exists* $M \in \mathfrak{gl}(\Lambda)$ *such that* $\|M\| \leq r^{1/2} |b|^{-1}$ *and* $\theta := M'b$ *satisfies* $|\theta| = 1$ *and belongs to* $\partial\Lambda$.

PROOF In terms of a spectral decomposition

$$b = \sum_{i=1}^{r} \mu_i c_i, \qquad |\mu_1| \geq \cdots \geq |\mu_r|$$

we have $|b| = (\Sigma \mu_i^2)^{1/2}$. Put $y := (1/\mu_1)\, c_1$ and $M := M_y \in \mathfrak{gl}(\Lambda)$. Then

$$\theta := M'b = y \circ b = c_1 \in \partial\Lambda$$

has norm 1 and $\|M\| = |y| = |\mu_1|^{-1} \leq r^{1/2} |b|^{-1}$. Q.E.D.

5

Geometric Differential Inequalities

In this section we estimate the derivatives of certain functions or transformations on Λ (or Π). Since the constants involved in these estimates usually depend on the order of the derivative taken, we write

$$\left|\left(\frac{\partial}{\partial t}\right)^{\alpha} f(t)\right| \leq C(\alpha) \cdot g(t) \tag{5.1}$$

for all $\alpha \in \mathbb{N}^n$ and $t \in \Lambda$, where the constant $C(\alpha)$ depends only on α and Λ but not on the point $t \in \Lambda$. Similar notations apply to the case where $f(t)$ is vector-valued and also to functions on Π (in which case the derivatives depend on multi-indices in $\mathbb{N}^{2n}$).

LEMMA 5.2
Consider the diffeomorphism

$$y \mapsto \varphi(y) := S_y e \tag{5.3}$$

of Λ *and its Jacobian matrix* $d\varphi(y) \in GL(n,\mathbb{R})$. *Then, for every multi-index* $\alpha \in \mathbb{N}^n$, *we have*

$$\left|\left(\frac{\partial}{\partial y}\right)^{\alpha} [d\varphi(y)^{-1}]\right| \leq C(\alpha)\ \delta(e,\varphi(y))^{1+|\alpha|} \tag{5.4}$$

PROOF Since $\varphi(y) = y^2$ in the Jordan algebraic sense, we have $d\varphi(y)h = 2y \circ h$ for all $h \in \mathbb{R}^n$, and, thus, $d\varphi(y) = 2M_y$. Let Φ: $GL(n,\mathbb{R}) \to GL(n,\mathbb{R})$ denote the inversion map $\Phi(A) := A^{-1}$. It has the derivative

$$d\Phi(A)B := -A^{-1}BA^{-1} \ (B \in \mathbb{R}^{n\times n}).$$

Now fix $k \in \mathbb{N}$ and consider the k-th order derivative

$$d^k\Phi(A)(B_1 \otimes \cdots \otimes B_k) = (-1)^k \sum_{\sigma \in \mathcal{S}_k} A^{-1}B_{\sigma(1)} A^{-1} \cdots A^{-1}B_{\sigma(k)}A^{-1}$$

for $B_1, \ldots, B_k \in \mathbb{R}^{n\times n}$. Here $\mathcal{S}_k$ is the permutation group. Since $y \mapsto M_y$ is linear, the mapping $\psi(y) := d\varphi(y)^{-1} = \Phi(2M_y)$ from Λ into $GL(n,\mathbb{R})$ has the k-th order derivative

$$d^k\psi(y)(x_1 \otimes \cdots \otimes x_k) = \frac{(-1)^k}{2^{k+1}} \sum_{\sigma \in \mathcal{S}_k} M_y^{-1}M_{x_{\sigma(1)}}M_y^{-1} \cdots M_y^{-1}M_{x_{\sigma(k)}}M_y^{-1}$$

for all $x_1, \ldots, x_k \in \mathbb{R}^n$. For the operator norm $\|\cdot\|$ on $\mathbb{R}^{n\times n}$ this implies

$$\begin{aligned}\|d^k\psi(y)(x_1 \otimes \cdots \otimes x_k)\| &\le C(k)\|M_y^{-1}\|^{k+1}\|M_{x_{\sigma(1)}}\| \cdots \|M_{x_{\sigma(k)}}\| \\ &\le C(k)\|M_y^{-1}\|^{k+1} \cdot |x_{\sigma(1)}| \cdots |x_{\sigma(k)}|.\end{aligned}$$

It follows that $d^k\psi(y)$ has norm

$$|||d^k\psi(y)||| \le C(k)\|M_y^{-1}\|^{k+1} \le C(k)\delta(e, y)^{2(k+1)} \le C(k)\delta(e, S_ye)^{k+1},$$

using (4.37) and Lemma 4.18. Q.E.D.

LEMMA 5.5
The diffeomorphism $t \mapsto S(t) = t^{-1}$ *of* Λ *satisfies the derivative estimates*

$$\left|\left(\frac{\partial}{\partial t}\right)^{\alpha} S(t)\right| \le C_\alpha |t^{-1}|^{1+|\alpha|}$$

for all multi-indices $\alpha \in \mathbb{N}^n$.

PROOF By (3.9), we have $dS(t) = -P_t^{-1}$ for all $t \in \Lambda$. Let

$$P(s,t) = \frac{1}{2}(P_{s+t} - P_s - P_t) = M_sM_t + M_tM_s - M_{s\circ t} \qquad (5.6)$$

be the bilinear polarization of the quadratic map $t \mapsto \varphi(t) := P_t$. Then an induction argument shows for every $k \in \mathbb{N}$

$$d^k(\varphi \circ S)(t)(x_1 \otimes \cdots \otimes x_k) = \sum_{I,J} P(d^{|I|}S(t) \bigotimes_{i\in I} x_i, d^{|J|}S(t) \bigotimes_{j\in J} x_j).$$

Here I and J form a partition of $\{1, \ldots, k\}$, into sets of order $|I|$ and $|J| = k - |I|$, respectively. Since $P(s,t)$ is bilinear, we obtain by induction

$$\begin{aligned} |d^{k+1}S(t)| &= |d^k(\varphi \circ S)(t)| \\ &\leq C(k) \sum_{I,J} |d^{|I|}S(t)||d^{|J|}S(t)| \\ &\leq C(k) \sum_{I,J} |t^{-1}|^{|I|+1}|t^{-1}|^{|J|+1} \\ &\leq C(k)|t^{-1}|^{k+2}. \end{aligned}$$

Q.E.D.

LEMMA 5.7
The determinant function $t \mapsto \Delta(t)$ *of* Λ *satisfies the derivative estimates*

$$\left|\left(\frac{\partial}{\partial t}\right)^{\alpha} \Delta(t)^{\lambda}\right| \leq C(\alpha, \lambda)\Delta(t)^{\lambda}|t^{-1}|^{|\alpha|}$$

for all multi-indices $\alpha \in \mathbb{N}^n$ *and* $\lambda \in \mathbb{R}$.

PROOF Identifying $\mathbb{R}^n$ with its dual space via the quadratic form (2.77), the function $t \mapsto \log \Delta(t)$ has the gradient

$$\text{grad}\,[\log \Delta(t)] = \frac{\text{grad}\,\Delta(t)}{\Delta(t)} = t^{-1}, \tag{5.8}$$

as follows from (2.43) and the fact that, in the irreducible case, ω is a power of Δ. Applying Lemma 5.5, it follows that

$$\left|\left(\frac{\partial}{\partial t}\right)^{\alpha} \log \Delta(t)\right| \leq C_\alpha\, |t^{-1}|^{|\alpha|} \tag{5.9}$$

for every $\alpha \in \mathbb{N}^n$. Lemma 5.7 follows in an elementary way. Q.E.D.

LEMMA 5.10
The function $(s,t) \mapsto \delta(s,t)$ on $\Lambda \times \Lambda$ *satisfies the derivative estimates*

$$\left|\left(\frac{\partial}{\partial s}\right)^{\alpha}\left(\frac{\partial}{\partial t}\right)^{\beta} \delta(s, t)\right| \leq C(\alpha, \beta)\delta(s, t)\delta(e, s)^{2|\alpha|}\delta(e, t)^{2|\beta|} \tag{5.11}$$

for all multi-indices $\alpha, \beta \in \mathbb{N}^n$.

PROOF Since $s + t > s$, (4.17) implies $(s + t)^{-1} < s^{-1}$. For the order-unit norm $\|\cdot\|$ on $\mathbb{R}^n$ (cf. [U12]) we obtain

$$\|(s + t)^{-1}\| \leq \min(\|s^{-1}\|, \|t^{-1}\|),$$

and, therefore, for the euclidean norm,

$$|(s + t)^{-1}| \lesssim \min(|s^{-1}|,|t^{-1}|) \lesssim \min(\delta(e,s)^2, \delta(e,t)^2). \tag{5.12}$$

Using the definition (4.6) of $\delta(s,t)$, we see that

$$\left(\frac{\partial}{\partial s}\right)^{\alpha}\left(\frac{\partial}{\partial t}\right)^{\beta}\delta(s, t) = 2^{-r}\left(\frac{\partial}{\partial s}\right)^{\alpha}\left(\frac{\partial}{\partial t}\right)^{\beta}[\Delta(s)^{-1/2}\Delta(t)^{-1/2}\Delta(s + t)]$$

is a linear combination of terms of the form

$$\left(\frac{\partial}{\partial s}\right)^{\alpha-\mu}\Delta(s)^{-1/2}\cdot\left(\frac{\partial}{\partial t}\right)^{\beta-\nu}\Delta(t)^{-1/2}\left(\frac{\partial}{\partial s}\right)^{\mu}\left(\frac{\partial}{\partial t}\right)^{\nu}\Delta(s + t). \tag{5.13}$$

Since $|s^{-1}| \lesssim \delta(e,s)^2$ by (4.37), Lemma 5.7 yields

$$\left|\left(\frac{\partial}{\partial s}\right)^{\alpha-\mu}\Delta(s)^{-1/2}\right| \leq C(\alpha)|\Delta(s)|^{-1/2}\cdot|s^{-1}|^{|\alpha|-|\mu|}$$

$$\leq C(\alpha)\,|\Delta(s)|^{-1/2}\cdot\delta(e,s)^{2(|\alpha|-|\mu|)}, \tag{5.14}$$

and, in a similar way,

$$\left|\left(\frac{\partial}{\partial t}\right)^{\beta-\nu}\Delta(t)^{-1/2}\right| \leq C(\beta)|\Delta(t)|^{-1/2}\cdot\delta(e, t)^{2(|\beta|-|\nu|)}, \tag{5.15}$$

and

$$\left|\left(\frac{\partial}{\partial s}\right)^{\mu}\left(\frac{\partial}{\partial t}\right)^{\nu}\Delta(s + t)\right| \leq C(\mu, \nu)\Delta(s + t)|(s + t)^{-1}|^{|\mu|+|\nu|} \tag{5.16}$$

$$\leq C(\mu, \nu)\,\Delta(s + t)\,\delta(e, s)^{2|\mu|}\,\delta(e, t)^{2|\nu|}.$$

Combining (5.14), (5.15), and (5.16), the assertion follows. Q.E.D.

LEMMA 5.17
For all w $\in \Pi$, *the complex derivatives of* $\delta(z,w)^2$ *satisfy*

$$\left|\left(\frac{\partial}{\partial z}\right)^{\alpha}\left(\frac{\partial}{\partial z^*}\right)^{\beta}\delta(z, w)^2_{z=e}\right| \leq C(\alpha, \beta)\,\delta(e, w)^2$$

for all multi-indices $\alpha,\beta \in \mathbb{N}^n$.

PROOF For every $z = x + i\xi \in \Pi$, the operator $P_{e+x}^{-1/2}$ acting on $\mathbb{C}^n$ is a contraction: $\|P_{e+x}^{-1/2}\| \leq 1$. The spectral decomposition (2.84) implies

$$|(e + iP_{e+x}^{-1/2}\xi)^{-1}| \lesssim 1.$$

It follows that

$$|(e + z)^{-1}| = |[P_{e+x}^{1/2}(e + iP_{e+x}^{-1/2}\xi)]^{-1}| = |P_{e+x}^{-1/2}(e + iP_{e+x}^{-1/2}\xi)^{-1}|$$
$$\leq \|P_{e+x}^{-1/2}\|\cdot|(e + iP_{e+x}^{-1/2}\xi)^{-1}| \lesssim 1. \quad (5.18)$$

The holomorphic function $z \mapsto \Delta(z)$ does not vanish on the domain Π, and the gradient formula

$$\operatorname{grad} \log \Delta(z) = z^{-1}, \quad (5.19)$$

generalizing (5.8), holds by analytic continuation. The explicit computation of the higher-order derivatives on Λ, made in the proof of Lemmas 5.5 and 5.7, carries over to Π and yields, for every multi-index $\alpha \in \mathbb{N}^n$ and $z \in \Pi$,

$$\left|\left(\frac{\partial}{\partial z}\right)^{\alpha}\Delta(z + w^*)_{z=e}\right| = \left|\left(\frac{\partial}{\partial w^*}\right)^{\alpha}\Delta(e + w^*)\right|$$
$$\leq C(\alpha)|\Delta(e + w^*)|\cdot|(e + w^*)^{-1}|^{|\alpha|} \leq C(\alpha)|\Delta(e + w^*)|. \quad (5.20)$$

As

$$\delta(z, w)^2 = \frac{\Delta(z + w^*)\overline{\Delta(z + w^*)}}{\Delta(z + z^*)\Delta(w + w^*)},$$

an application of Leibniz' rule together with (5.20) proves Lemma 5.17.

Q.E.D.

LEMMA 5.21
For some N_0, *one has*

$$\int_\Lambda \delta(e, y)^{-N_0}\, dm(y) < \infty$$

and

$$\int_\Pi \delta(e, z)^{-N_0}\, d\mu(z) < \infty.$$

where $d\mu$ *denotes any Aut*(Π)*-invariant measure on* Π [*cf. (8.6)*].

PROOF One may perform an orthogonal change of coordinates so that (assuming that Λ is irreducible without loss of generality) the determinant function Δ does not vanish identically along any of the coordinate axes: this means that, for all j, y_j times a nonzero constant is a derivative of $\Delta(y)$. By (5.20), one has $|y_j| \leq C\, \Delta(e + y)$, which implies that, for N_0 large enough, the first integral is finite since, as $\Delta(y) \leq \Delta(e + y)$, one has

$$\int_\Lambda \left[\frac{\Delta(e + y)}{\Delta(y)^{1/2}}\right]^{-N_0} dm(y) \leq \int_\Lambda (\Delta(e + y))^{(-1/2)(N_0 - 2n/r)} dy. \tag{5.22}$$

The same goes for the second integral by virtue of (5.20). Q.E.D.

In the complex irreducible case, one may use the properties of the reproducing kernel of the λ-Bergman space (cf. Section 14) to check that $N_0 > 2(p - 1)$ is sufficient (cf. [UU; (3.14) and (1.21)]), if p is the genus of Π [cf. (4.55)].

6

Weights and Classes of Symbols

Let Λ be a symmetric cone in $\mathbb{R}^n$, not necessarily irreducible, and consider the cotangent bundle

$$T^*\Lambda \approx \Lambda \times \mathbb{R}^n$$

and the complex tube domain

$$\Pi = \{z \in \mathbb{C}^n : \mathrm{Re}(z) \in \Lambda\}.$$

Define a diffeomorphism $\tau : \Pi \to T^*\Lambda$ by putting

$$\tau(x + i\xi) := (x^{-1}, -\xi) \tag{6.1}$$

where $x^{-1} = Sx$ is the *symmetry* of Λ around e. Now consider the semi-direct product $G = GL(\Lambda) \times \mathbb{R}^n$ of all pairs $\gamma = (P,b)$, with the actions

$$[\gamma](y,\eta) = [P,b](y,\eta) := (Py, P'^{-1}\eta + b) \tag{6.2}$$

on $T^*\Lambda$ and

$$\gamma \cdot z = (P,b) \cdot z := P'^{-1}z - ib \tag{6.3}$$

on Π. It follows from the formula $(Py)^{-1} = P'^{-1}y^{-1}$ proved in (2.12) that τ intertwines these actions.

If m is a function on $T^*\Lambda$ we define

$$\tilde{m} = m \circ \tau. \tag{6.4}$$

DEFINITION 6.5 Given $C_1 > 0$ and $N_1 \geq 0$, a positive measurable function m on $T^*\Lambda$ is called a *weight function of type* (C_1, N_1) if

$$\tilde{m}(z) \le C_1\, \tilde{m}(w)\, e^{N_1 d(z,w)} \tag{6.6}$$

for all (z,w) $\in \Pi \times \Pi$. In view of (4.51), an equivalent condition is

$$\tilde{m}(z) \le C_1\, \tilde{m}(w)\, \delta(z,w)^{N_1} \tag{6.7}$$

with a possibly different pair (C_1, N_1). In terms of m, this may be written as

$$m(x,\xi) \le C_1\, m(y,\eta)\, [\delta(x,y) + \min(|\xi - \eta|_x, |\xi - \eta|_y)]^{N_1} \tag{6.8}$$

for all (x,ξ) and $(y,\eta) \in T^*\Lambda$: indeed, this is again a consequence of (4.51) together with the fact that if $y = Pe$, so that $Sy = P'^{-1}e$, one has

$$\|\xi - \eta\|_{Sy} = |P'(\xi - \eta)| = |\xi - \eta|_y. \tag{6.9}$$

It will be essential, in some parts of this work, to get estimates not depending on weight-functions, only on their type.

LEMMA 6.10
The product of two weight functions is a weight function and so are arbitrary real powers of a weight function. The following functions are weight functions

$$m(y, \eta) := \omega(y) \tag{6.11}$$

$$m_0(y, \eta) := (1 + |\eta|_y^2)^{1/2}. \tag{6.12}$$

Here ω *is the inverse density (3.27) of the GL(Λ)-invariant measure with respect to the Lebesgue measure.*

PROOF The first two assertions are trivial. We now claim that

$$1 + \log \omega(s) \le C\,(1 + \log \delta(s,e)) \tag{6.13}$$

for all $s \in \Lambda$. To see this we may assume that Λ is irreducible of rank r. Then

$$\omega(s) = \Delta(s)^{n/r} = \left(\prod_{j=1}^{r} \lambda_j\right)^{1/2}$$

using the spectral decomposition (4.10) of s. Comparing with (4.11), the assertion (6.13) follows. By GL(Λ)-invariance of $\omega(x)/\omega(y)$ and δ we obtain the estimate (6.8) for $m(y,\eta) := \omega(y)$. In order to prove that (6.12) is a weight function, we note that by Peetre's inequality

$$(1 + |\eta|_y^2)^{1/2} \le 2^{1/2}(1 + |\xi - \eta|_y^2)^{1/2}(1 + |\xi|_y^2)^{1/2},$$

valid for (x,ξ) and (y,η) in $T^*\Lambda$, it suffices to show that

$$|\xi|_y \le C\, |\xi|_z \cdot \delta(y,z)^2, \tag{6.14}$$

which was proved in (4.30). Q.E.D.

DEFINITION 6.15 Let $f \in C^\infty(T^*\Lambda)$ and $(y,\eta) \in T^*\Lambda$. For any pair (p,q) of nonnegative integers, let $\|f\|_{p,q;\,y,\eta}$ denote the smallest constant such that the (p,q)-th derivative satisfies

$$|d^{p,q}f(y,\eta)(e_1 \otimes \cdots \otimes e_p, \epsilon_1 \otimes \cdots \otimes \epsilon_q)| \le \|f\|_{p,q;\,y,\eta}\|e_1\|_y \cdots \|e_p\|_y |\epsilon_1|_y \cdots |\epsilon_q|_y. \tag{6.16}$$

In particular

$$\|f\|_{0,0;\,y,\eta} = |f(y,\eta)|. \tag{6.17}$$

REMARK
For short, one may think of an operator $\Sigma\, b_j\, \partial/\partial y_j$ (resp: $\Sigma\, \beta_j\, \partial/\partial \eta_j$) with $\|b\|_y \le C$ (resp. $|\beta|_y \le C$) as a differentiation operator bounded at y.

LEMMA 6.18
The semi-norms (6.16) are G-*invariant in the sense that we have for each* $\gamma \in$ G *and* $f \in C^\infty(T^*\Lambda)$

$$\|f \circ [\gamma]\|_{p,q;\,y,\eta} = \|f\|_{p,q;[\gamma](y,\eta)}. \tag{6.19}$$

PROOF By (6.2) we have for $\gamma = (P,\xi)$

$$d^{p,q}(f \circ [\gamma])(y,\eta)(e_1 \otimes \cdots \otimes e_p, \epsilon_1 \otimes \cdots \otimes \epsilon_q)$$
$$= (d^{p,q}f)(Py, P'^{-1}\eta + \xi)\,(Pe_1 \otimes \cdots \otimes Pe_p, P'^{-1}\epsilon_1 \otimes \cdots \otimes P'^{-1}\epsilon_q).$$

Since $\|e_i\|_y = \|Pe_i\|_{Py}$ and $|\epsilon_j|_y = |P'^{-1}\epsilon_j|_{Py}$, the assertion follows. Q.E.D.

As a special case of (6.21) we obtain

$$\|f\|_{p,q;\,y,\eta} = \|f \circ [\gamma]\|_{p,q;\,e,0} \tag{6.20}$$

if $[\gamma](e,0) = (y,\eta)$.

DEFINITION 6.21 Let m be a weight function on $T^*\Lambda$. A smooth function f on $T^*\Lambda$ is called a *symbol of uniform type* and *weight* m if for each (p,q) we have

$$\|f\|_{p,q;\,y,\eta} \leq C(p,q) \cdot m(y,\eta)$$

uniformly for $(y,\eta) \in T^*\Lambda$. For each nonnegative integer N we put

$$|||f|||_{m,N} := \sum_{p+q\leq N} \sup \{m(y,\eta)^{-1}\|f\|_{p,q;\,y,\eta}\}. \tag{6.22}$$

In the space of all symbols of uniform type and weight m, we say that a sequence (f_k) *converges* to f if $f = \lim f_k$ in the C^∞-topology on $C^\infty(T^*\Lambda)$ and, in addition, for any N, the sequence $|||f_k|||_{m,N}$ is bounded.

REMARK 6.23
1) We shall use freely, if the need arises, the obvious notion of symbol of weight m up to the order of differentiability N; 2) the notion of convergence used is the one suitable for future applications of Lebesgue's dominated convergence theorem; the following essential fact would not be true if the topology defined by the semi-norms (6.22) were the one used on the space of symbols.

PROPOSITION 6.24
Any symbol of uniform type and weight m *is the limit, in that class of symbols, of a sequence of smooth symbols with compact supports.*

PROOF It is identical to that of Proposition 6.8 in [U6]. It is clear that $m(y,\eta) = \delta(e,\tau^{-1}(y,\eta))$ is a weight function, and Lemma 5.17, together with (4.51) and (6.22), shows that m is a symbol of weight m. Let χ be a smooth function on $\mathbb{R}_+$, with $\chi(t) = 1$ for $t \leq 1$ and $\chi(t) = 0$ for $t \geq 2$. Then, one may approximate any symbol f by

$$f_k(y,\eta) = f(y,\eta)\, \chi(k^{-1}\delta(e,\tau^{-1}(y,\eta))). \tag{6.25}$$

Q.E.D.

We have seen in Section 0 that, in the pseudodifferential analysis on

$\mathbb{R}^n$, it was necessary to introduce, besides symbols of uniform type, classes of symbols within which asymptotic expansions can be performed in an easy way. The same situation arises in the present context.

DEFINITION 6.26 Let m_0 be the weight function defined in Lemma 6.10. Let k be a real number. A *symbol* of *classical type* and *order* k is any smooth function f on $T^*\Lambda$ such that, for every pair of nonnegative integers (p,q), one has, with some constant C(p,q), the estimate

$$\|f\|_{p,q;\,y,\eta} \leq C(p,q)\,(m_0(y,\eta))^{k-q}. \tag{6.27}$$

Symbols of classical type but of a more general weight shall be needed later (Definition 12.1). We shall denote as S^k the space of symbols of classical type and order k.

7

The Family of μ-Symbols

DEFINITION 7.1 Let f be a smooth symbol with compact support and let $\mu \in \mathbb{R}$. We denote as a_μ the function on $T^*\Lambda$ characterized by the identity $(s,t \in \Lambda)$

$$(\mathcal{F}_2^{-1} a_\mu)(s, s - t) = \delta(s,t)^\mu (\mathcal{F}_2^{-1} a)(s, s - t) \tag{7.2}$$

in which a is associated to f by the formula (1.32). Let us recall that $(\mathcal{F}_2^{-1} a)(s,t)$ is zero unless $s \in \Lambda$ and $s - t \in \Lambda$ and that, as a consequence, the same holds for a_μ.

PROPOSITION 7.3
The symbols f *and* a_μ *are linked by the following two formulas:*

$$a_\mu(s, \xi) = 2^n \int_{T^*\Lambda} f(y, \eta) e^{2i\pi\langle \eta - \xi, s - S_y s\rangle} \delta(s, S_y s)^\mu \, dy \, d\eta \tag{7.4}$$

and

$$f(y, \eta) = 2^{-n} \int_{T^*\Lambda} a_\mu(s, \xi) e^{2i\pi\langle \xi - \eta, s - S_y s\rangle} \delta(s, S_y s)^{-\mu}$$
$$\mathrm{Det}\left(\frac{\partial}{\partial y}(S_y s)\right) \det\left(\frac{\partial}{\partial s}(s - S_y s)\right) ds \, d\xi. \tag{7.5}$$

PROOF The proof is that of the formulas (1.32) and (1.35): one just has to plug in an extra factor $\delta(s,t)^{\mu}$ and perform the change of variable $t \mapsto y = \mathrm{mid}(s,t)$ in the first integral. Q.E.D.

LEMMA 7.6
Let the operator A *have a* μ*-symbol* a_{μ} *and let* $u \in C_0^{\infty}(\Lambda)$. *For every* $s \in \Lambda$, *let*

$$v_s(t) = \delta(s,t)^{-\mu} u(t), \tag{7.7}$$

and let $\hat{v}_s$ *denote the Fourier transform of* v_s. *Then*

$$(Au)(s) = \int_{\mathbb{R}^n} a_{\mu}(s, \xi) e^{2i\pi\langle \xi, s\rangle}\, \hat{v}_s(\xi)\, d\xi. \tag{7.8}$$

PROOF From (1.30) and Definition 7.1, we get

$$(Au)(s) = \int_{\mathbb{R}^n} (\mathcal{F}_2^{-1} a_{\mu})(s, s - t)\, \delta(s, t)^{-\mu}\, u(t)\, dt \tag{7.9}$$

from which the formula follows. Q.E.D.

The interest of μ-symbols lies in this formula, since $\hat{v}_s(\xi)$ can be very nice for large μ.

THEOREM 7.10
For any real number μ, *the map* $f \mapsto a_{\mu}$ *defined by the integral formula (7.4) in the case when* $f \in C_0^{\infty}(T^*\Lambda)$ *extends as a sequentially continuous map from the space of symbols of uniform type and any given weight* m *to itself. More precisely, assume that* m *is of type* (C_1,N_1), *i.e., satisfies*

$$\tilde{m}(z) \leq C_1\, \tilde{m}(w)\, \delta(z,w)^{N_1} \tag{7.11}$$

for some $C_1 > 0$ and $N_1 \geq 0$, *and let* N *be any nonnegative integer: then, with some* $\tilde{N} = \tilde{N}(\mu,N,N_1)$ *and* $C = C(\mu,N,N_1,C_1)$, *one has*

$$|||a_{\mu}|||_{m,N} \leq C |||f|||_{m,\tilde{N}}. \tag{7.12}$$

PROOF Observe first that the estimate does not depend on the weight function m, except for C_1 and N_1 that occur in (7.11)—this uniformity shall be crucial at one point. If $(P,b) \in G$, $a_{\mu} \circ [P,b]$ is the function associated, under the same map (7.4), to $f \circ [P,b]$. To prove this, one can either show that the μ-symbol is G-covariant, or notice that $S_y P = P S_{P^{-1}y}$ and perform, in the integral, the measure-preserving change of coordinates defined by $(y,\eta) = [P,b](y',\eta')$. Using (6.22), we see that it is sufficient to prove the required estimates for a_μ and its derivatives at $(s,\xi) = (e,0)$. The advantage of this is that the differentiation operators are normalized there in the standard norm. We need to perform integrations by parts with the help of vector fields bounded at y in the sense of the remark following Definition 6.15. It is easier, of course, to deal with operators with constant coefficients. To this effect, fix a smooth nonnegative function h on $\mathbb{R}^+$, supported in [0,2], such that

$$\int_\Lambda h(\delta(e,x))\, dx = 1. \tag{7.13}$$

Then

$$\omega(y)^{-1} \int_\Lambda h(\delta(x,y))\, dx = 1, \tag{7.14}$$

and

$$a_\mu(s,\xi) = \int_\Lambda a_\mu^x(s,\xi)\, dx, \tag{7.15}$$

with

$$a_\mu^x(s,\xi) \tag{7.16}$$

$$= 2^n \int_{T^*\Lambda} e^{2i\pi\langle \eta-\xi, s-S_y s\rangle}\, \delta(s, S_y s)^\mu\, f(y,\eta)\, \omega(y)^{-1}\, h(\delta(x,y))\, dy\, d\eta.$$

Fix $P \in GL(\Lambda)$ with $x = Pe$. Then,

$$a_\mu^x(Ps, P'^{-1}\xi) = 2^n \int_{T^*\Lambda} e^{2i\pi\langle \eta-\xi, s-S_y s\rangle}\, \delta(s, S_y s)^\mu$$

$$f(Py, P'^{-1}\eta)\, (\omega(x)\omega(y))^{-1}\, h(\delta(e,y))\, dy\, d\eta. \tag{7.17}$$

After a first integration by parts, it becomes

$$a_\mu^x(Ps, P'^{-1}\xi) = 2^n\omega(x)^{-1}\int_{T^*\Lambda} e^{2i\pi\langle\eta-\xi, s-S_ys\rangle}\,\delta(s, S_ys)^\mu\, h(\delta(e, y))$$

$$(1 + |s - S_ys|^2)^{-N}\left[1 - (2\pi)^{-2}\sum_j \frac{\partial^2}{\partial\eta_j^2}\right]^N f(Py, P'^{-1}\eta)\,\omega(y)^{-1}\,dy\,d\eta. \quad (7.18)$$

Also, with $t = S_ys$, consider the differential operator

$$L_j = -\frac{1}{2i\pi}\frac{\partial}{\partial t_j} = -\frac{1}{2i\pi}\sum_k \frac{\partial y_k}{\partial t_j}\frac{\partial}{\partial y_k} \quad (7.19)$$

so that

$$e^{2i\pi\langle\eta-\xi, s-S_ys\rangle}(\eta_j - \xi_j) = L_j(e^{2i\pi\langle\eta-\xi, s-S_ys\rangle}). \quad (7.20)$$

As L_j is formally self-adjoint with respect to the measure $dt = d(S_ys)$ on Λ, a new integration by parts permits one to write

$$a_\mu^x(Ps, P'^{-1}\xi) = 2^n\omega(x)^{-1}\int_{T^*\Lambda} (1 + |\eta - \xi|^2)^{-k} e^{2i\pi\langle\eta-\xi, s-S_ys\rangle}$$

$$\left(1 + \sum_j L_j^2\right)^k F(s, y, \eta)\,\mathrm{Det}\left(\frac{\partial}{\partial y}(S_ys)\right) dy\,d\eta, \quad (7.21)$$

with

$$F(s, y, \eta) = \delta(s, S_ys)^\mu\, h(\delta(e, y))\,(1 + |s - S_ys|^2)^{-N}$$

$$\omega(y)^{-1}\mathrm{Det}\left(\frac{\partial}{\partial y}(S_ys)\right)^{-1}\left[1 - (2\pi)^{-2}\sum_j \frac{\partial^2}{\partial\eta_j^2}\right]^N f(Py, P'^{-1}\eta). \quad (7.22)$$

We must now make a careful estimate of the derivatives involved. First, we estimate the size of the coefficients of

$$L^\alpha = (-2i\pi)^{-|\alpha|}\left(\frac{\partial}{\partial t}\right)^\alpha \quad (7.23)$$

when expressed as linear combinations of $(\partial/\partial y)^\beta$. Let $s = Qe$ for $Q \in GL(\Lambda)$, and put $y = Qy'$, $t = Qt'$. Since $t' = S_{y'}e$, Lemma 5.2 implies

$$\left|\left(\frac{\partial}{\partial y'}\right)^{\alpha}\left(\frac{\partial y'_k}{\partial t'_j}\right)\right| \leq C(\alpha)\delta(e, t')^{1+|\alpha|}. \tag{7.24}$$

Since $\|Q\| \lesssim \delta(e,s)^2$ by Lemma 4.29, we obtain

$$\left|\left(\frac{\partial}{\partial y}\right)^{\alpha}\left(\frac{\partial y_k}{\partial t_j}\right)\right| \leq C(\alpha)\ \delta(s, S_y s)^{1+|\alpha|}\ \delta(e, s)^{2(2+|\alpha|)}. \tag{7.25}$$

Therefore, a product of a number $k_1 \leq 2k$ of operators L_j is a linear combination of operators $(\partial/\partial y)^{\alpha}$ with $|\alpha| \leq 2k$, the coefficients being less than $C(k)\ (\delta(s,S_y s)\ \delta(e,s)^4)^{2k}$. Another consequence of (7.25) is

$$\left|\left(\frac{\partial}{\partial y}\right)^{\alpha} \operatorname{Det}\left(\frac{\partial}{\partial y}(S_y s)\right)^{-1}\right| = \left|\left(\frac{\partial}{\partial y}\right)^{\alpha} \operatorname{Det}\left(\frac{\partial}{\partial y'}(S_{y'} e)\right)^{-1}\right|$$

$$\leq C(\alpha)\ \delta(e, t')^{n+|\alpha|}\ \delta(e, s)^{2|\alpha|} = C(\alpha)\ \delta(s, S_y s)^{n+|\alpha|}\ \delta(e, s)^{2|\alpha|}. \tag{7.26}$$

When applying L_j to $\delta(s,S_y s)^{\mu} = \delta(s,t)^{\mu}$, we may use Lemma 5.10:

$$\left|\left(\frac{\partial}{\partial t}\right)^{\alpha} \delta(s, t)^{\mu}\right| \leq C(\alpha, \mu)\ \delta(s, t)^{\mu}\ \delta(e, t)^{2|\alpha|}$$

$$\leq C(\alpha, \mu)\ \delta(s, S_y s)^{\mu+2|\alpha|}\ \delta(e, s)^{2|\alpha|}. \tag{7.27}$$

Of course, applying $L_j = (-\ 1/2\pi i)\ \partial/\partial t_j$ to $(1 + |s - S_y s|^2)^{-N} = (1 + |s - t|^2)^{-N}$ is harmless, and the $\partial/\partial y$-derivatives of $h(\delta(e,y))$ and $\omega(y)^{-1}$ are bounded when $\delta(e,y) \leq 2$. Finally, when $\delta(e,y) \leq 2$, (6.22) and (4.30) show that

$$\left|\left(\frac{\partial}{\partial y}\right)^{\alpha}\left(\frac{\partial}{\partial \eta}\right)^{\beta} f(Py, P'^{-1}\eta)\right| \leq C(\alpha, \beta)\ m(Py, P'^{-1}\eta)\ |||f|||_{m,|\alpha+\beta|}. \tag{7.28}$$

Also, by (6.8) and (4.23),

$$m(Py, P'^{-1}\eta) \leq C_1\ m(Ps, P'^{-1}\xi)\ \delta(s, y)^{N_1}\ (1 + |\eta - \xi|_y)^{N_1}. \tag{7.29}$$

We have finished our examination of the derivatives of the various factors involved. It remains to be noted that

$$\operatorname{Det}\left(\frac{\partial}{\partial y}(S_y s)\right) = \operatorname{Det}\left(\frac{\partial}{\partial y'}(S_{y'} e)\right) = 2^n \operatorname{Det} M_{y'}, \tag{7.30}$$

where $M_{y'}$ is the operator of multiplication by y' in a Jordan algebra so that

$$\left|\operatorname{Det}\left(\frac{\partial}{\partial y}(S_y s)\right)\right| \leq C\,|y'|^n \leq C\,\delta(e, y')^{2n} = C\,\delta(s, y)^{2n}. \qquad (7.31)$$

Starting from (7.21) we obtain

$$|a_\mu^x(Ps, P'^{-1}\xi)| \leq C_1\, C(\mu, k, N)\, \omega(x)^{-1} |||f|||_{m,2k+2N}$$

$$m(Ps, P'^{-1}\xi) \int (1 + |\xi - \eta|)^{N_1-2k} (1 + |s - S_y s|^2)^{-N}\, \delta(e, s)^{12k}$$

$$\delta(s, S_y s)^{\mu+6k+n} \delta(s, y)^{N_1+2n}\, dy\, d\eta, \qquad (7.32)$$

where the integral is performed where $\delta(e,y) \leq 2$. Assuming k large enough so that $N_1 - 2k < -n$, using Lemma 4.18 to the effect that $\delta(s,S_y s)$ is of the same size as $\delta(s,y)^2$ and using the inequality

$$1 + |s - S_y s|^2 \geq C^{-1}\, \delta(s,S_y s)^{1/N_0} \qquad (7.33)$$

proved in Lemma 4.27, we find after an obvious change of variable and recalling that $x = Pe$,

$$|a_\mu^x(s, \xi)| \leq C(\mu, k, N, C_1, N_1)\, \omega(x)^{-2} |||f|||_{m,2k+2N}\, m(s, \xi)$$
$$\delta(x, s)^{12k} \int_{\delta(x,y)\leq 2} \delta(s, y)^{2\mu+4n+12k+N_1-(2N/N_0)}\, dy. \qquad (7.34)$$

When $\delta(x,y) \leq 2$, one has $\delta(x,s) \leq C\,\delta(s,y)$, $\omega(x) \geq C^{-1}\,\omega(y)$ and

$$\int_{\delta(x,y)\leq 2} dx \leq C\,\omega(y). \qquad (7.35)$$

Thus, starting from (7.15) and (7.34), we obtain

$$|a_\mu(s, \xi)| \leq C(\mu, k, N, C_1, N_1)\, |||f|||_{m,2k+2N}\, m(s, \xi)$$
$$\int_\Lambda \delta(s, y)^{2\mu+4n+24k+N_1-(2N/N_0)} dm(y), \qquad (7.36)$$

which is just the estimate of $|a_\mu|$ that we need in view of Lemma 5.21. Also, from Lebesgue's dominated convergence theorem, it is clear that a_μ depends on f in a sequentially continuous way if we use for a_μ only the notion of convergence that applies to symbols up to the order of differentiability zero (cf. Remark 6.23).

We have to examine now the derivatives of a_μ. Starting from the $\partial/\partial\xi$-derivatives, what we have to do is to examine the derivatives of

$$(s - S_y s)^\alpha \, e^{2i\pi\langle s - S_y s, \eta - \xi\rangle} \, \delta(s, S_y s)^\mu$$

with respect to s, at s = e. The first factor on the left is taken care of by Lemma 4.27, at the price of having to change μ to $\mu + 2|\alpha|$ in the estimate (7.36); also, since $|\eta - \xi|_s \le C\,|\eta - \xi|_y\, \delta(s,y)^2$, and using Lemma 5.10 again, it is clear that what is left is to control the derivatives of $S_y s$ with respect to s, at s = e. From Lemma 5.5 and the usual change of coordinates y = Pe, s = Ps′ it follows that

$$\left|\left(\frac{\partial}{\partial s}\right)^\alpha (S_y s)\right| \le C(\alpha)\, \delta(s,y)^{2+2|\alpha|}\, |Ss'|^{1+|\alpha|} \le C(\alpha)\, \delta(s,y)^{3+3|\alpha|} \quad (7.37)$$

Q.E.D.

8

Coherent States

Let Λ be a symmetric cone in $\mathbb{R}^n$, with right half-space

$$\Pi := \{z \in \mathbb{C}^n : \mathrm{Re}(z) \in \Lambda\}. \tag{8.1}$$

A set of coherent states is a family of functions in $L^2(\Lambda)$ from which one can build a resolution of the identity. First, we define the coherent states associated with a certain weighted Laplace transformation on Λ (depending on some real parameter λ). Though somewhat overspecialized (it will need to be generalized soon), this family of functions plays a natural role in connection with the λ-Bergman space discussed in Section 14. Let Δ be the determinant function of Λ. In the case when $\Lambda = \Pi\Lambda_i$ is a product of irreducible cones, we have

$$\Delta(t) = \prod_i \Delta_i(t_i) \tag{8.2}$$

where Δ_i is the i-th Jordan algebraic determinant function. Recall [(3.27) and (3.42)] that

$$\omega(t) := \prod_i \Delta_i(t_i)^{n_i/r_i}. \tag{8.3}$$

DEFINITION 8.4 Let λ be a real parameter. For every $z \in \Pi$, the function ψ_z^λ on Λ, defined by

$$\psi_z^\lambda(t) := \Delta(\mathrm{Re}z)^{\lambda/2}\, \Delta(t)^{\lambda/2}\, e^{-2\pi\langle t,z\rangle} \tag{8.5}$$

is called the *coherent state* at z. Here $\langle t,z\rangle$ is the standard bilinear form on $\mathbb{C}^n \times \mathbb{C}^n$. Let

$$d\mu(z) = \omega(\mathrm{Re}\, z)^{-2}\, d(\mathrm{Re}\, z)\, d(\mathrm{Im}\, z) \tag{8.6}$$

be the Aut(Π)-invariant measure on Π, where Aut (Π) was defined at the end of Section 2.

PROPOSITION 8.7
There exists a parameter λ_0 such that for all $\lambda > \lambda_0$ we have for all $z \in \Pi$

$$\psi_z^\lambda \in L^2(\Lambda),$$

and moreover

$$\int_\Pi |(u|\psi_z^\lambda)|^2\, d\mu(z) = c_\lambda\, \|u\|^2 \tag{8.8}$$

for every $u \in L^2(\Lambda)$. Here c_λ is a finite constant.

PROOF We may assume that Λ is irreducible of rank r. The integral

$$I_\lambda = \int_\Lambda \Delta(y)^\lambda\, e^{-4\pi\langle e,y\rangle}\, dm(y) = (4\pi)^{-\lambda r}\, \Gamma_\Lambda(\lambda) \tag{8.9}$$

is finite for $\lambda > n/r - 1$. Here Γ_Λ is the generalized Γ-function introduced by M. Koecher [via (8.9)] and expressed in terms of the classical Γ-function by S. Gindikin [cf.(14.7) and [G1]]. Now write $x = P'^{-1}e$ for some $P \in GL(\Lambda)$. Since

$$\Delta(Py) = \Delta(Pe)\, \Delta(y) \tag{8.10}$$

it follows that

$$\begin{aligned}
&\Delta(x)^\lambda \int_\Lambda \Delta(t)^\lambda e^{-4\pi\langle t,x\rangle}\, dm(t)\\
&= \Delta(x)^\lambda \int_\Lambda \Delta(Py)^\lambda\, e^{-4\pi\langle y,P'x\rangle}\, dm(y)\\
&= \Delta(x)^\lambda \Delta(Pe)^\lambda \int_\Lambda \Delta(y)^\lambda\, e^{-4\pi\langle y,e\rangle}\, dm(y) = I_\lambda
\end{aligned}$$

is independent of x. Write $z = x + i\xi$. Then

$$\int_\Lambda |\psi_z^\lambda(t)|^2 \, dm(t) = \int_\Lambda \Delta(x)^\lambda \, \Delta(t)^\lambda \, e^{-4\pi\langle t,x\rangle} \, dm(t) = I_\lambda$$

implies that $\psi_z^\lambda \in L^2(\Lambda)$ for all $\lambda > n/r - 1$. Now let $u \in L^2(\Lambda)$. Applying Parseval's formula to the ξ-variable we obtain, with $p = 2n/r$,

$$\int_\Pi |(u|\psi_z^\lambda)|^2 \, d\mu(z) = \int_\Pi \Delta(x)^{-p} \, dx \, d\xi$$

$$\left| \int_\Lambda \overline{u(t)} \, \Delta(x)^{\lambda/2} \, \Delta(t)^{(\lambda-p)/2} \, e^{-2\pi\langle t,x\rangle} \, e^{-2i\pi\langle t,\xi\rangle} dt \right|^2$$

$$= \int_\Lambda \Delta(x)^{\lambda-p} \int_\Lambda |u(t)|^2 \Delta(t)^{\lambda-p} \, e^{-4\pi\langle t,x\rangle} \, dt \, dx$$

$$= \int_\Lambda |u(t)|^2 \, \Delta(t)^{\lambda-p/2} \int_\Lambda \Delta(x)^{\lambda-p/2} \, e^{-4\pi\langle t,x\rangle} \, dm(x) \, dm(t)$$

$$= I_{\lambda-p/2} \int_\Lambda |u(t)|^2 \, dm(t) = I_{\lambda-p/2} \|u\|^2,$$

where $I_{\lambda-p/2} < \infty$ if $\lambda > p - 1$. Q.E.D.

PROPOSITION 8.11
With U *defined by (3.24) we have*

$$U(\dot{\gamma})\psi_z^\lambda = \psi_{\gamma\cdot z}^\lambda \tag{8.12}$$

for all $\gamma = (P,b) \in G$ *and* $z \in \Pi$. *Recall that the action of* G *on* Π *has been defined in* (6.3).

PROOF

$$U(\gamma)\psi_z^\lambda(t) = \psi_z^\lambda(P^{-1}t) \, e^{2\pi i\langle t,b\rangle}$$

$$= \Delta(x)^{\lambda/2} \, \Delta(P^{-1}t)^{\lambda/2} \, e^{-2\pi\langle P^{-1}t, x+i\xi\rangle} \, e^{2i\pi\langle t,b\rangle}$$

$$= \Delta(P'^{-1}x)^{\lambda/2} \, \Delta(t)^{\lambda/2} \, e^{-2\pi\langle t, P'^{-1}z - ib\rangle}.$$

Q.E.D.

DEFINITION 8.13 Let $\mathscr{A}$ be the algebra of operators on $L^2(\Lambda)$ generated by differential operators with polynomial coefficients and by the multiplication by $\Delta(t)^{-1}$. We let $\mathscr{S}(\Lambda)$ denote the space of $u \in C^\infty(\Lambda)$ such that Au

$\in L^2(\Lambda)$ for every $A \in \mathcal{A}$. Giving $\mathcal{S}(\Lambda)$ its natural (Fréchet) topology, we let $\mathcal{S}'(\Lambda)$ denote the topological dual of $\mathcal{S}(\Lambda)$.

REMARKS 8.14

1) With a reference to the decomposition of Λ as a product of irreducible cones and the corresponding formula (8.2), it is clear that the multiplication by powers, with arbitrary real exponents, of any $\Delta_j(t_j)$, preserves the space $\mathcal{S}(\Lambda)$; this will be used in a moment. As another consequence, even though $\mathcal{A}$ is not stable under the operation of taking the adjoint, $\mathcal{S}'(\Lambda)$ may be described as the set of linear combinations of distributions Au with $u \in L^2(\Lambda)$ and $A \in \mathcal{A}$. 2) Given any finite-dimensional linear subspace E of $\mathcal{A}$, containing the scalars, one may define $\mathcal{S}_E(\Lambda)$ by the property that $u \in \mathcal{S}_E(\Lambda)$ if and only if $Au \in L^2(\Lambda)$ for every $A \in E$: then any $v \in \mathcal{S}'(\Lambda)$ extends as a continuous linear form on $\mathcal{S}_E(\Lambda)$ if E is a well-chosen subspace. The functions ψ_z^λ defined by (8.5) do not lie in $\mathcal{S}(\Lambda)$; however, given any finite-dimensional E, they lie in $\mathcal{S}_E(\Lambda)$ if λ is large enough: this gives a meaning to $(\psi_z^\lambda | v)$ for any fixed $v \in \mathcal{S}'(\Lambda)$ when λ is large.

THEOREM 8.15

For every distribution $u \in \mathcal{S}'(\Lambda)$ *one has*

$$I_{\nu,\lambda} = \int_\Pi e^{2\nu d(e,z)}\, |(u|\psi_z^\lambda)|^2\, d\mu(z) < \infty$$

if $-\nu$ *and* λ *are large enough. The distribution* u *belongs to* $L^2(\Lambda)$ *if and only if* $I_{0,\lambda} < \infty$ *for* λ *large enough, and it belongs to* $\mathcal{S}(\Lambda)$ *if and only if, for every* $\nu \in \mathbb{N}$, *one has* $I_{\nu,\lambda} < \infty$ *for* λ *large enough.*

PROOF As in the proof of [U6; Theorem 10.2] we put

$$(\psi_z^\lambda | u) = \int_\Lambda u(t)\, \Delta(x)^{\lambda/2}\, \Delta(t)^{(\lambda-p)/2}\, e^{-2\pi\langle z^*,t\rangle}\, dt \tag{8.16}$$

and use integration by parts to obtain

$$2\pi z_j^*\, (\psi_z^\lambda | u) = -\int_\Lambda u(t)\, \Delta(x)^{\lambda/2}\, \Delta(t)^{(\lambda-p)/2}\, \frac{\partial}{\partial t_j} e^{-2\pi\langle z^*,t\rangle}\, dt$$

$$= \int_\Lambda \frac{\partial}{\partial t_j} [u(t)\, \Delta(t)^{(\lambda-p)/2}]\, \Delta(x)^{\lambda/2}\, e^{-2\pi\langle z^*,t\rangle}\, dt$$

$$= \left(\psi_z^\lambda \middle| \frac{\partial u}{\partial t_j}\right) + \frac{\lambda - p}{2}\left(\psi_z^\lambda \,|\, \Delta^{-1}\frac{\partial \Delta}{\partial t_j}\, u\right), \tag{8.17}$$

assuming, for simplicity of notations only, that Λ is irreducible and setting $p = 2n/r$. Since

$$\Delta(x)^{-1/2}\,(\psi_z^\lambda | u) = (\psi_z^{\lambda-1} | \Delta(t)^{1/2} u), \tag{8.18}$$

it follows that

$$2^{-r}\,\frac{\Delta(e + z^*)}{\Delta(x)^{1/2}}\,(\psi_z^\lambda | u) = (\psi_z^{\lambda-1} | B_\lambda u) \tag{8.19}$$

for some operator B_λ lying in the algebra generated by differential operators with polynomial coefficients and by the operator of multiplication by $\Delta(t)^{-1/2}$. Now, from Definition (4.43), the absolute value of the factor in front of the left-hand side is just $\delta(e,z)$, which should be compared to $e^{d(e,z)}$ by Lemma 4.50. In view of (8.8) and (8.19), what we have proved up to now is that $I_{0,\lambda} < \infty$ for large λ if $u \in L^2(\Lambda)$ and that, given ν, one has $I_{\nu,\lambda} < \infty$ for λ large enough if $u \in \mathcal{S}(\Lambda)$. To prove the rest of Theorem 8.15, we need a lemma.

LEMMA 8.20

Let $\nu \in \mathbb{R}$ be given. There exists $C(\nu)$ such that, for every function f holomorphic on Π, one has

$$\sum_k \int_\Pi \left|\frac{\partial f}{\partial z_k}\right|^2 e^{2\nu d(e,z)}\, d\mu(z)$$

$$\leq C(\nu) \int_\Pi |f(z)|^2\, e^{2\nu d(e,z)}\, \delta(e, z)^4\, d\mu(z). \tag{8.21}$$

PROOF It rests on the estimate

$$\delta(z,z') \leq c_0\, \delta(z,w)\, \delta(w,z') \tag{8.22}$$

valid for any three points of Π, and proved (in the irreducible case, with $c_0 = 2^r$) in [UU; Lemma 3.18] and on Lemma 4.61; a second look at the proof of this latter lemma allows to rephrase it as follows. There exists $c_0 > 0$ such that, for any $z \in \Pi$ with $z = x + iy$ and any $z' \in \mathbb{C}^n$, the inequality $|z - z'| \leq c_0^{-1}\, \delta(e,x)^{-2}$ implies that $z' \in \Pi$ and $\delta(z,z') \leq c_0$.

Also recall Lemma 4.46: if $z = x + i\xi$ and $z' = x' + i\xi' \in \Pi$, one has $\delta(z,z') \geq \delta(x,x')$. Given $z \in \Pi$, set

$$\rho(z) = c_0^{-1}\, n^{-1/2}\, \delta(e,z)^{-2} \tag{8.23}$$

and consider the polydisk D_z in $\mathbb{C}^n$ centered at z with radii $\rho(z)$. According to what precedes and to Lemma 4.50, $|d(e,z) - d(e,z')| \leq d(z,z')$ remains bounded, in a uniform way, as $z' \in D_z$. Also, Cauchy's formula and the Cauchy-Schwarz inequality permit one to write

$$\left|\frac{\partial f}{\partial z_k}\right|^2 \leq 2\pi^{-n}\, \rho(z)^{-2n-2} \int_{D_z} |f(z')|^2\, dx'\, d\xi'. \tag{8.24}$$

In the irreducible case, one has $d\mu(z) = \Delta(x)^{-p}\, dx\, d\xi$ with $p = 2n/r$. With the help of Lemma 4.33, one can see that $\Delta(x')/\Delta(x)$ remains bounded in a uniform way as z' describes D_z. Thus, in order to complete the proof of Lemma 8.20, all that remains to be verified is the elementary fact that

$$\int_{|z-z'|\leq\rho(z)} \rho(z)^{-2n-2}\, dx\, d\xi \leq C\, \delta(e,z)^4. \tag{8.25}$$

End of proof of Theorem 8.15. Start from the identity

$$\int_\Pi (u|\psi_z^\lambda)(\psi_z^\lambda|v)\, d\mu(z) = c_\lambda\, (u|v) \tag{8.26}$$

valid for all $u,v \in L^2(\Lambda)$, if λ is large enough, as a consequence of (8.8). Now let B be any operator lying in the algebra defined right after (8.19). We then claim that the integral

$$\int_\Pi (u|\psi_z^\lambda)\, (\psi_z^\lambda|Bw)\, d\mu(z), \tag{8.27}$$

considered as a sesquilinear form on the pair (u,w), extends, if λ is large enough, as a continuous form on $L^2(\Lambda) \times \mathcal{S}(\Lambda)$. Indeed, we may let B, or rather its adjoint B^*, act on ψ_z^λ rather than w and observe that the function $\chi_z^\lambda(t) = \Delta(x)^{-\lambda/2}\, \psi_z^\lambda(t)$ depends on z in a holomorphic way and that

$$t_k\, \chi_z^\lambda(t) = -(2\pi)^{-1} \frac{\partial}{\partial z_k} \chi_z^\lambda(t). \tag{8.28}$$

Together with (8.18) and the part already proven of Theorem 8.15, Lemma

8.20 implies the claim just made. As a consequence, given $v \in \mathcal{S}'(\Lambda)$, there exists λ_0 such that, for all $\lambda > \lambda_0$ and all $u \in \mathcal{S}(\Lambda)$, (8.25) is still valid. The end of the proof of Theorem 8.15 is then a straightforward matter. Q.E.D.

9

From Symbols to Operators: The Main Estimate and Continuity

As the coherent states introduced in Definition 8.4 do not lie in $\mathcal{S}(\Lambda)$, we need to generalize this latter notion somewhat.

DEFINITION 9.1 Given $\lambda \in \mathbb{R}$, the space $\mathcal{S}_\lambda(\Lambda)$ is the space of all C^∞ functions φ on Λ such that for all nonnegative integers k and N, the k-th derivative satisfies

$$|d^k\varphi(t)(e_1 \otimes \cdots \otimes e_k)| \leq C(k, N)\, \Delta(t)^{\lambda/2}\, \Delta(e + t)^{-N}\, \|e_1\|_t \cdots \|e_k\|_t \tag{9.2}$$

for every $t \in \Lambda$ and all vectors $e_1, \ldots, e_k$.

LEMMA 9.3
Given $\varphi \in \mathcal{S}_\lambda(\Lambda)$, $\mu \in \mathbb{R}$ and $P \in GL(\Lambda)$ *define for all* $(s,\xi) \in \Lambda \times \mathbb{R}^n$

$$\chi_{\mu,P}(s, \xi) = \int_\Lambda \varphi(P^{-1}t)\, e^{-2\pi i\langle\xi,t\rangle}\, \delta(s, t)^{-\mu}\, dt. \tag{9.4}$$

Then, for all nonnegative integers p, q, ℓ *the* (p,q)*-th derivative in* (s,ξ) *satisfies*

$$\begin{aligned} &|d^{p,q}\chi_{\mu,P}(s, \xi)(e_1 \otimes \cdots \otimes e_p, \epsilon_1 \otimes \cdots \otimes \epsilon_q)| \\ &\quad\leq C(\varphi, \mu, p, q, \ell)\, \omega(Pe)\, \delta(s, Pe)^{-\mu+2p} \\ &\quad\quad (1 + |\xi|^2_{Pe})^{-\ell}\|e_1\|_{Pe} \cdots \|e_p\|_{Pe}\, |\epsilon_1|_{Pe} \cdots |\epsilon_q|_{Pe} \end{aligned} \tag{9.5}$$

provided $\lambda > \lambda_0(\mu, \ell)$.

PROOF Since

$$\chi_{\mu,P}(s,\xi) = \omega(Pe)\ \chi_{\mu,I}(P^{-1}s,P'\xi) \tag{9.6}$$

and

$$\|e_j\|_{Pe} = |P^{-1}e_j|, \quad |\epsilon_j|_{Pe} = |P'\epsilon_j|, \quad |\xi|_{Pe} = |P'\xi|, \tag{9.7}$$

the estimate can be reduced to the case when $P = I$, which we assume from now on. With

$$\chi_{\mu,I}(s,\xi) = \int_\Lambda \varphi(t)\ e^{-2i\pi\langle\xi,t\rangle}\ \delta(s,t)^{-\mu}\ dt, \tag{9.8}$$

we wish to prove that

$$\left|\left(\frac{\partial}{\partial s}\right)^\alpha \left(\frac{\partial}{\partial \xi}\right)^\beta \chi_{\mu,I}(s,\xi)\right| \leq C(\varphi,\mu,\alpha,\beta,\ell)\ \delta(e,s)^{-\mu+2|\alpha|}\ (1+|\xi|^2)^{-\ell} \tag{9.9}$$

provided that $\lambda > \lambda_0(\mu,\ell)$. With arbitrary large ℓ_1, one has

$$|\varphi(t)| \leq C(\varphi,\ell_1)\ \Delta(t)^{\lambda/2}\ \Delta(e+t)^{-\ell_1}. \tag{9.10}$$

Using the fact that

$$\Delta(t)^{1/2} = 2^{-r}\ \Delta(e+t)\ \delta(e,t)^{-1}, \tag{9.11}$$

we get from (9.8) (if $\lambda > 0$)

$$|\chi_{\mu,I}(s,\xi)| \leq C(\varphi,\ell_1) \int_\Lambda \delta(e,t)^{-\lambda/2}\ \Delta(e+t)^{\lambda/2-\ell_1}\ \delta(s,t)^{-\mu}\ dt \tag{9.12}$$

and, using Lemma 4.12, and assuming $\ell_1 \geq \lambda/2$,

$$|\chi_{\mu,I}(s,\xi)| \leq C(\varphi,\mu,\ell_1)\ \delta(e,s)^{-\mu} \int_\Lambda \delta(e,t)^{|\mu|-\lambda/2}\ dt, \tag{9.13}$$

where the integral is finite if $\lambda > \lambda_0(\mu)$ by Lemma 5.21. To save the required power of $1+|\xi|^2$, it is clear that we have to perform at most 2ℓ integrations by parts in (9.8): the operators $\partial/\partial t_j$ to be used are normalized at e, not at t, so, by Lemma 4.29, we may lose $C_0\ \delta(e,t)^2$ each time; Lemma 5.10

shows that the same loss may occur when we differentiate powers of $\delta(s,t)$. Overall, the worst possible loss in the integral is $\delta(e,t)^{4\ell}$, so that, from (9.13), we see that it can be controlled by an increase in λ. Finally, we must take $\partial/\partial s$ or $\partial/\partial\xi$ -derivatives in (9.8): as $|t| \leq C\,\Delta(e+t)$ (using a spectral resolution of t), the second species can be dealt with at no harm; for the first kind, Lemma 5.10 shows that we may lose $\delta(e,s)^2$ each time, which concludes the proof of the lemma. Q.E.D.

REMARKS 9.14
1) If $u(t) = \varphi(P^{-1}t)$, $\chi_{\mu,P}(s,\xi)$ is just $\hat{v}_s(\xi)$ as defined in Lemma 7.6, which accounts for the importance of that function; 2) in the case when

$$\varphi(t) = \Delta(t)^{\lambda/2}\, e^{-2\pi\langle e,t\rangle}, \tag{9.15}$$

one may write, using Definition 8.4,

$$\chi_{\mu,P}(s,\xi) = \int_\Lambda \psi_z^\lambda(t)\, \delta(s,t)^{-\mu}\, dt \tag{9.16}$$

with $z = P'^{-1}e + i\xi$; one may check that φ, as just defined, does belong to $\mathcal{S}_\lambda(\Lambda)$. Indeed, the condition (9.2) may be rephrased as

$$|(\frac{\partial}{\partial s})^\alpha(\varphi(Ps))(e)| \leq C(\alpha,N)\, \Delta(Pe)^{\lambda/2}\, \Delta(e+Pe)^{-N}. \tag{9.17}$$

As

$$\varphi(Ps) = (\Delta(Pe)\Delta(s))^{\lambda/2}\, e^{-2\pi\langle P'e,s\rangle}, \tag{9.18}$$

the estimate is a consequence of the inequality $|P'e| \leq C\,\langle P'e,e\rangle = C\,\langle e,Pe\rangle$, that follows from the self-adjointness of the cone Λ.

THEOREM 9.19
Let $A = Op(f)$, *where* f *is a symbol of uniform type of weight* m, *and the weight* m *is of type* (C_1,N_1). *Let* $\varphi \in \mathcal{S}_\lambda(\Lambda)$ *and* $N \geq 0$. *Then one has the estimate*

$$(U(\gamma)\varphi\,|\,AU(\gamma_1)\varphi) \leq C(C_1, N_1, \varphi, N)\, |||f|||_{m,\tilde{N}}\, \tilde{m}(\gamma\cdot e)\, e^{-Nd(\gamma_1\cdot e,\gamma\cdot e)} \tag{9.20}$$

for all $\gamma, \gamma_1 \in G$, *provided* $\tilde{N}$ *is larger than some* $\tilde{N}_0\,(N,N_1)$ *and* λ *is larger than some* $\lambda_0(N,N_1)$.

PROOF Let $\gamma = (P, -\xi)$, $\gamma_1 = (P_1, -\xi_1)$, $u = U(\gamma)\varphi$ and $u_1 = U(\gamma_1)\varphi$. From Lemma 7.6, it follows that

$$(u_1 \mid Au) = \int_\Lambda \bar{u}_1(s)\, dm(s) \int_{T^*\Lambda} a_\mu(s, \eta)\, e^{2i\pi\langle\eta, s-t\rangle}$$
$$u(t)\, \delta(s, t)^{-\mu}\, dt\, d\eta \tag{9.21}$$

if a_μ is the μ-symbol of A. From Theorem 7.10 we may assume that a_μ is a symbol of weight m for all μ, leaving to the reader the soft analysis argument based on the sequential continuity of the map $f \mapsto a_\mu$ and on the approximation property asserted by Proposition 6.24. Now put $z = \gamma \cdot e = x + i\xi$, $z_1 = \gamma_1 \cdot e = x_1 + i\xi_1$, and, in particular, $x = P'^{-1}e$, $x_1 = P_1'^{-1}e$. Then, since $u = U(\gamma)\varphi$,

$$\int_\Lambda e^{-2i\pi\langle\eta, t\rangle}\, u(t)\, \delta(s, t)^{-\mu}\, dt = \chi_{\mu,P}(s, \xi + \eta) \tag{9.22}$$

in the sense of (9.4). Therefore,

$$(u_1 \mid Au) = \int_{\mathbb{R}^n} \int_\Lambda \bar{u}_1(s)\, a_\mu(s, \eta)\, e^{2i\pi\langle\eta, s\rangle}\, \chi_{\mu,P}(s, \xi + \eta)\, dm(s)\, d\eta. \tag{9.23}$$

For all ℓ, one has (since $Pe = Sx$)

$$|\chi_{\mu,P}(s, \xi + \eta)| \le C(\varphi, \mu, \ell)\, \omega(x)^{-1}\, \delta(s, Sx)^{-\mu}\, (1 + \|\xi + \eta\|_x^2)u^{-\ell} \tag{9.24}$$

if $\lambda_0 > \lambda_0(\mu, \ell)$. Also, from Definition 9.1, one has for all ℓ_1 the estimate

$$|\varphi(t)| \le C(\varphi, \ell_1, \lambda)\, \Delta(t)^{\lambda/2}\, \Delta(e + t)^{-\ell_1}. \tag{9.25}$$

Finally, from (6.4), (6.8), and (6.9) we obtain

$$m(s, \eta) \le C_1\, \tilde{m}(z)\, \delta(s, Sx)^{N_1}\, (1 + \|\xi + \eta\|_x^2)^{N_1}. \tag{9.26}$$

Thus, choosing ℓ first so that $2\ell > 2N_1 + n$, we get from (9.23) (changing η to $P'^{-1}\,\eta - \xi$ and performing the $d\eta$-integration at once)

$$|(u_1 | Au)| \leq C(\varphi, \mu, C_1, N_1, \ell_1, \lambda) \, |||a_\mu|||_{m,0} \, \tilde{m}(z)$$
$$\int_\Lambda \Delta(P_1^{-1}s)^{\lambda/2} \, \Delta(e + P_1^{-1}s)^{-\ell_1} \, \delta(s, Sx)^{-\mu+N_1} \frac{ds}{\omega(s)}. \quad (9.27)$$

We note, by Lemma 6.10, that

$$\omega(s)^{-1} \leq C_0 \, \omega(x) \, \delta(s, Sx)^{N_0} \quad (9.28)$$

for some N_0 depending only on Λ and that the integral

$$I = \int_\Lambda \Delta(P_1^{-1}s)^{\lambda/2} \, \Delta(e + P_1^{-1}s)^{-\ell_1} \, \delta(s, Sx)^{-\mu+N_1+N_0} \, ds \quad (9.29)$$

would be just $\chi_{\mu-N_1-N_0,P_1}(Sx,0)$ if, in Lemma 9.3, φ were replaced by

$$\psi(t) = \Delta(t)^{\lambda/2}\Delta(e + t)^{-\ell_1}. \quad (9.30)$$

Although ψ does not belong to $\mathcal{S}_\lambda(\Lambda)$, it remains true that, for any finite set of continuous semi-norms on $\mathcal{S}_\lambda(\Lambda)$, ψ does belong to the Banach space of functions characterized by the finiteness of these semi-norms provided that ℓ_1 is large enough. Thus Lemma 9.3 applies and yields

$$|I| \leq C(\mu - N_1 - N_0, \lambda, \ell_1) \, \omega(x_1)^{-1} \, \delta(Sx, Sx_1)^{-\mu+N_0+N_1} \quad (9.31)$$

if λ is greater than some λ_0 depending only on $\mu - N_0 - N_1$ (and on n, Λ, but this is a standing assumption) and ℓ_1 is greater than some number depending only on λ and on $\mu - N_0 - N_1$. Finally, as

$$\omega(x) \, \omega(x_1)^{-1} \leq C_0 \, \delta(x,x_1)^{N_0}, \quad (9.32)$$

we get

$$|(u_1|Au)| \leq C(\varphi,\mu,C_1,N_1,\ell_1,\lambda) \, |||a_\mu|||_{m,0} \, \tilde{m}(z) \, \delta(x,x_1)^{-\mu+2N_0+N_1} \quad (9.33)$$

and note that we have to choose μ,λ,ℓ_1 in this order. In view of Theorem 7.10, this is the required estimate (since μ may be chosen arbitrarily large) provided we still show how to control powers of $1 + \|\xi - \xi_1\|_x^2$. This is done, starting from (9.23), with the help of integrations by parts with respect to $\partial/\partial s$, more specifically using operators $\Sigma\alpha_j \, \partial/\partial s_j$ normalized at $Pe = Sx$, i.e., satisfying $\|\alpha\|_{Sx} = 1$. In this way, making $\overline{u_1}(s)$ explicit as

$$\overline{u_1}(s) = \bar{\varphi}(P_1^{-1}s)\, e^{2i\pi\langle\xi_1,s\rangle}, \tag{9.34}$$

we can save powers of $(1 + |\xi_1 + \eta|^2_{Sx})^{-1} = (1 + \|\xi_1 + \eta\|^2_x)^{-1}$. As, on the other hand, powers of $(1 + \|\xi + \eta\|^2_x)^{-1}$ may be saved by an increase of ℓ, we are done, provided we check the size of the derivatives involved. Each $\Sigma\, \alpha_j\, \partial/\partial s_j$ may be applied to $a_\mu\,(s,\eta)$ at a worst possible loss (Lemma 4.29) of $\delta(s,Sx)^2$ since they are $\partial/\partial s$-operators normalized at Sx, not s; the action of $\Sigma\, \alpha_j\, \partial/\partial s_j$ on $\chi_{\mu,P}(s, \xi + \eta)$ is controlled with a loss of the same size by Lemma 9.3; finally, from (6.22) and Definition 9.1, one sees that operators $\Sigma\, \alpha_j\, \partial/\partial s_j$ may be applied to $\varphi(P_1^{-1}s)$ at no loss when α is normalized at s, thus at a loss of at most $\delta(s,Sx)^2$ in the case under scrutiny. Since, in (9.29), μ can be taken arbitrarily large, the proof is complete. Q.E.D.

REMARK 9.35

When φ is given by (9.15), $U(P, -\xi)\varphi$ is just ψ_z^λ with $z = P'^{-1}\, e + i\xi = (P,-\xi)\, e$.

COROLLARY 9.36

If f *is a symbol of uniform type and weight* m, *then* Op(f) *extends as a continuous operator from* $\mathcal{S}(\Lambda)$ *to* $\mathcal{S}(\Lambda)$. *If the weight* m *is a constant, say* 1, *then* Op(f) *extends as a continuous operator from* $L^2(\Lambda)$ *to* $L^2(\Lambda)$: *moreover, for some* N_0 *and* C *depending only on* Λ *the norm of* Op(f) *on* $L^2(\Lambda)$ *is at most* $C|||f|||_{1,N_0}$.

PROOF In view of Theorem 8.15 it suffices, for the second part, to show that if f is a symbol of weight 1 then, for λ large enough,

$$k(z,z') = |(\psi_z^\lambda|\mathrm{Op}(f)\, \psi_{z'}^\lambda)| \tag{9.37}$$

is the kernel of some bounded operator from $L^2(\Pi, d\mu(z))$ to itself. In view of Remark 9.35, this follows from Theorem 9.19 and the second part of Lemma 5.21. For the first part, it is just the same, except that we must substitute for $L^2(\Pi)$ the space $L^2(\Pi, e^{2\nu d(e,z)}\, d\mu(z))$; since we have on one side of the matter our own choice of ν we are finished, using (6.6) with $w = e$. Q.E.D.

10

From Operators to Symbols: The Converse of the Main Estimate

THEOREM 10.1
Let A *be a continuous linear operator from* $\mathcal{S}(\Lambda)$ *to* $\mathcal{S}(\Lambda)$, *and let* m *be a weight function of type* (C_1, N_1). *Assume that for every number* N *and parameter* λ *there exists* $\varphi \in \mathcal{S}_\lambda(\Lambda)$, $\varphi \neq 0$, *such that for some constant* $C > 0$,

$$|(U(\gamma_1)\varphi|A\,U(\gamma)\varphi)| \leq C\,\tilde{m}(\gamma_1 \cdot e)\,e^{-Nd(\gamma\cdot e,\gamma_1\cdot e)} \tag{10.2}$$

for all $\gamma,\gamma_1 \in G$. *Call* $|||A|||_{m,\varphi,N}$ *the least constant* C *such that the inequality above holds. Then, for all* $\mu \in \mathbb{R}$, A *has a* μ*-symbol of uniform type and weight* m *in the sense of Definition 7.1. Moreover, for all* N, μ, *and* λ *large enough,*

$$|||a_\mu|||_{m,N} \leq C(\varphi,\mu,N,N_1,C_1)\,|||A|||_{m,\varphi,\tilde{N}} \tag{10.3}$$

for some $\tilde{N}$ *depending only on* (μ,N_1,N).

First we need a lemma.

LEMMA 10.4
Let $d\gamma$ *be a left Haar measure on* G. *If* $\lambda > \lambda_0$ *for some* λ_0 *depending only on* Λ, *then for every* $\varphi \in \mathcal{S}_\lambda(\Lambda)$ *there exists a finite constant* $c(\varphi)$ *such that*

$$\int_G |(U(\gamma)\varphi|u)|^2\,d\gamma = c(\varphi)\,\|u\|^2 \tag{10.5}$$

for all $u \in L^2(\Lambda)$.

PROOF The elements $\gamma = (P, -\xi)$ of G can be characterized by the pairs $(z,k) \in \Pi \times O(\Lambda)$ [where $O(\Lambda)$ was defined in (2.5)] such that

$$z := x + i\xi := P'^{-1}e + i\xi$$

and

$$P'^{-1} = P_x^{1/2}k. \tag{10.6}$$

Then, we claim that $d\gamma = d\mu(z)\, dk$, where $d\mu(z)$ was defined in (8.6) and dk is the (normalized) Haar measure on $O(\Lambda)$, is a left Haar measure on G. Indeed, let $\gamma_1 = (P_1, -\xi_1)$ and let $\gamma_2 = \gamma_1\gamma = (P_1P, -\xi_1 - P'^{-1}_1\xi)$ be associated with the pair (z_2, k_2), with

$$z_2 := x_2 + i\xi_2 = \gamma_1 z = P_1'^{-1}z + i\xi_1, \tag{10.7}$$

so that

$$P_1'^{-1}P'^{-1} = P_{x_2}^{1/2}k_2. \tag{10.8}$$

Since $x_2 = P'^{-1}_1 x$ and

$$P_{P_1'^{-1}x} = P_1'^{-1}P_xP_1^{-1} \tag{10.9}$$

by (2.9), one has

$$h := P_{x_2}^{-1/2}P_1'^{-1}P_x^{1/2} \in O(\Lambda) \tag{10.10}$$

(compute hh′) and

$$k_2 = hk. \tag{10.11}$$

Now h depends only on the pair (γ_1, z), not on k; moreover, $d\mu$ is a G-invariant measure on Π: hence (Fubini's theorem)

$$d\mu(z_2)\, dk_2 = d\mu(z)\, dk, \tag{10.12}$$

which shows that $d\gamma$ is, indeed, a left Haar measure on G. With $\gamma = (P, -\xi) = (P_x^{-1/2}k, -\xi)$ and with the help of Parseval's identity, one may write

$$\int_G |(U(\gamma)\varphi|u)|^2 \, d\gamma = \int_\Lambda |u(t)|^2 \, \omega(t)^{-2} \, I(t) \, dt \tag{10.13}$$

with

$$I(t) = \int_{\Lambda \times O(\Lambda)} |\varphi(k^{-1}P_x^{1/2}t)|^2 \, \omega(x)^{-2} \, dx \, dk. \tag{10.14}$$

Let $t = P_1^{-1}e$ with $P_1 \in GL(\Lambda)$, so that $P_x^{1/2} P_1^{-1} = h^{-1} P^{1/2}_{P'^{-1}_1 x}$ with the same h as in (10.10): thus $P_x^{1/2}t = h^{-1} P'^{-1}_1 x$. Performing the change of variable $x = P'_1 y$ and using the fact that dk is a Haar measure on $O(\Lambda)$, one gets for large λ

$$I(t) = \omega(t) \int_{\Lambda \times O(\Lambda)} |\varphi(k^{-1}y)|^2 \, \omega(y)^{-2} \, dy \, dk < \infty, \tag{10.15}$$

from which the lemma follows. Q.E.D.

Proof of Theorem 10.1. Given $\varphi \in \mathcal{S}_\lambda(\Lambda)$ with λ large enough, there exists $\psi \in \mathcal{S}(\Lambda)$ such that

$$u = \int_G (U(\gamma)\psi|u) \, U(\gamma)\varphi \, d\gamma \tag{10.16}$$

for all $u \in L^2(\Lambda)$. This is a consequence of Lemma 10.4 and of Schwarz's inequality together with Schur's lemma; that the integral does not vanish identically for all u, for some choice of $\psi \in \mathcal{S}(\Lambda)$ (which can then be renormalized), is a consequence of the fact that ψ may be chosen as an approximation of φ. Also, (10.16) remains true if φ and ψ are exchanged. If u and $v \in \mathcal{S}(\Lambda)$, one gets, after two applications of (10.16), and with $\gamma = (P,-\xi)$ and $\gamma_1 = (P_1,-\xi_1)$

$$(v|Au) = \int_{G \times G} (U(P,-\xi)\psi|u) \, K(P,\xi; P_1,\xi_1) \, (v|U(P_1,-\xi_1)\psi) \, d\gamma \, d\gamma_1 \tag{10.17}$$

with

$$K(P,\xi; P_1,\xi_1) = (U(P_1,-\xi_1)\varphi|AU(P,-\xi)\varphi). \tag{10.18}$$

From (1.30) and the relationship between a_μ and a given in (7.2), we see that a rank one operator $u \mapsto (\psi|u)\varphi$ has, for all μ, a μ-symbol a_μ characterized by

$$(\mathcal{F}_2^{-1}a_\mu)(s,s-t) = \delta(s,t)^\mu\ \varphi(s)\ \bar{\psi}\ (t)\ \omega(t)^{-1} \tag{10.19}$$

if s,t $\in \Lambda$ ($\mathcal{F}_2^{-1}a_\mu$ is zero on (s,θ) unless $s - \theta \in \Lambda$) so that

$$a_\mu(s,\eta) = \int_\Lambda e^{-2i\pi\langle\eta,s-t\rangle}\ \delta(s,t)^\mu\ \varphi(s)\ \bar{\psi}\ (t)\ \omega(t)^{-1}\ dt. \tag{10.20}$$

From (10.17) the operator A appears as a superposition of rank one operators so we have to study the μ-symbol

$$a_\mu(s,\eta) = \int_{\Lambda\times G\times G} e^{-2i\pi\langle\eta,s-t\rangle}\ \delta(s,t)^\mu\ K(P,\xi;P_1,\xi_1)\ \omega(t)^{-1}$$
$$\psi(P^{-1}s)\ e^{-2i\pi\langle\xi,s\rangle}\ \overline{\psi}(P_1^{-1}t)\ e^{2i\pi\langle\xi_1,t\rangle}\ dt\ d\gamma\ d\gamma_1. \tag{10.21}$$

We proceed to show that, when K satisfies the hypotheses of Theorem 10.1, the integral is convergent and that the symbol a_μ so defined is of uniform type and weight m. We leave to the reader the soft analysis argument required to show, then, that a_μ is indeed the μ-symbol of A. Let

$$\chi_{-\mu,P_1}(s,\xi_1+\eta) = \int_\Lambda \psi(P_1^{-1}t)\ e^{-2i\pi\langle\xi_1+\eta,t\rangle}\ \delta(s,t)^\mu\ [\omega(P_1'^{-1}e)\omega(t)]^{-1}\ dt. \tag{10.22}$$

This notation is consistent with (9.4) if $\varphi(t)$ is replaced there by $\omega(t)^{-1}\ \psi(t)$. Since $\psi \in \mathcal{S}(\Lambda)$, we may apply the estimates in Lemma 9.3 without even worrying about any λ. The integral in (10.21) becomes

$$a_\mu(s,\eta) = \int_{G\times G} e^{-2i\pi\langle\eta+\xi,s\rangle}\ K(P,\xi;P_1,-\xi_1)\ \omega(P_1'^{-1}e)$$
$$\psi(P^{-1}s)\ \overline{\chi}_{-\mu,P_1}(s,\xi_1+\eta)\ d\gamma\ d\gamma_1. \tag{10.23}$$

Let $z := x + i\xi := \gamma \cdot e$ and $z_1 := x_1 + i\xi_1 := \gamma_1 \cdot e$, so that $P'^{-1}_1 e = x_1$. We now majorize $|a_\mu(s,\eta)|$, using the estimate

$$|K(P,\xi;P_1,\xi_1)|$$
$$\leq C(N)\ \tilde{m}(z_1)\ |||A|||_{m,\varphi,N_0N}\ \delta(x,x_1)^{-N}\ (1+||\xi-\xi_1||_{x_1}^2)^{-N}, \tag{10.24}$$

a consequence of the assumption of Theorem 10.1 if, in accordance with Lemma 4.50, N_0 is chosen so that

$$\delta(x, x_1)(1 + \|\xi - \xi_1\|_{x_1}^2) \le C e^{N_0 d(z,z_1)} \tag{10.25}$$

for some $C > 0$ and all $z, z_1 \in \Pi$, as well as of the estimates

$$|\psi(P^{-1}s)| \le C(\lambda', \psi, \ell_1)\,(\Delta(x)\Delta(s))^{\lambda'}\,\Delta(e + P^{-1}s)^{-\ell_1} \tag{10.26}$$

(from Definition 9.1, with any choice of λ' or ℓ_1) and

$$|\chi_{-\mu,P_1}(s, \xi_1 + \eta)| \le C(\psi, \mu, \ell)\,\omega(x_1)^{-1}\,\delta(s, Sx_1)^{\mu} \\ (1 + \|\xi_1 + \eta\|_{x_1}^2)^{-\ell} \tag{10.27}$$

from Lemma 9.3. Also, we substitute in (10.24) the estimate (6.8)

$$\tilde{m}(z_1) \le C_1\, m(s, \eta)\, \delta(s, Sx_1)^{N_1}\,(1 + \|\xi_1 + \eta\|_{x_1}^2)^{N_1}. \tag{10.28}$$

Since all the right-hand sides in the last four estimates are right K-invariant as functions of γ and γ_1, we may dispense with the $dk\, dk_1$ integration and replace $d\gamma\, d\gamma_1$ by $(\omega(x)\omega(x_1))^{-2}\, dx\, dx_1\, d\xi\, d\xi_1$. We first note that

$$\int_{R^n \times R^n} (1 + \|\xi_1 + \eta\|_{x_1}^2)^{N_1 - \ell}(1 + \|\xi - \xi_1\|_{x_1}^2)^{-N}\, d\xi\, d\xi_1 \\ \le C(N, N_1)\,\omega(x_1)^2 \tag{10.29}$$

if we choose some ℓ, N with $\ell > N_1 + n/2$ and $N > n/2$. Then

$$|a_\mu(s, \eta)| \le C_1\, C(\lambda', \mu, \psi, N, \ell_1)\, m(s, \eta)\, |||A|||_{m,\varphi,N_0 N}\, I \tag{10.30}$$

with (N, ℓ_1) still to be chosen as large as we please, λ' exactly as we please and

$$I = \int_{\Lambda \times \Lambda} \delta(s, Sx_1)^{N_1 + \mu}\, \delta(x, x_1)^{-N} (\Delta(x)\Delta(s))^{\lambda'} \\ \Delta(e + P^{-1}s)^{-\ell_1}\, \omega(x)^{-2}\, dx\, dx_1. \tag{10.31}$$

Recall from Definition 4.6 that $\Delta(e + P^{-1}s) = 2^r\,(\Delta(x)\Delta(s))^{1/2}\,\delta(e,P^{-1}s)$ and that $\delta(e,P^{-1}s) = \delta(Pe,s) = \delta(Sx,s)$: choosing $\lambda' = \ell_1/2$, we get

$$|I| \le C(\ell_1) \int_{\Lambda \times \Lambda} \delta(s, Sx_1)^{N_1 + \mu}\, \delta(s, Sx)^{-\ell_1}\, \delta(x, x_1)^{-N} \omega(x)^{-2}\, dx\, dx_1 \tag{10.32}$$

with N and ℓ_1 still to be chosen. As

$$\delta(s, Sx_1) \leq C\, \delta(s, Sx)\, \delta(x, x_1) \tag{10.33}$$

and $\omega(x)^{-2} \leq C\, (\omega(x)\omega(x_1))^{-1}\, \delta(x,x_1)^{N_0}$, for some N_0 depending only on Λ, we finally get

$$|I| \leq C(\ell_1, N_1, \mu) \int_{\Lambda\times\Lambda} \delta(x, x_1)^{N_0+N_1+\mu-N}$$
$$\delta(Ss, x)^{N_1+\mu-\ell_1} dm(x)\, dm(x_1), \tag{10.34}$$

a bounded integral if N and ℓ_1 are large enough in view of Lemma 5.21 [and GL(Λ)-invariance of δ as well as dm]. This gives the required estimate (10.30) for $|a_\mu|$ (observe that in the statement of Theorem 10.1 we have made C depend on φ, not ψ; this is so because only φ is given, and we may choose $\psi \in \mathcal{S}(\Lambda)$ as we please except for a normalization that depends on φ). We finally have to majorize the derivatives of a_μ, starting from (10.23). Using covariance, we may study the derivatives of $a_\mu(s,\eta)$ only at $s = e$. The $\partial/\partial\eta_j$-derivatives are taken care of quite easily since, by Lemma 9.3, no loss is incurred when letting $\partial/\partial s_j$-differentiation operators, normalized at Sx_1, act on $\chi_{-\mu,P_1}(s, \xi_1 + \eta)$. Thus, at each time, we may lose at most $C\, \delta(e,Sx_1)^2$, a factor under control since we have been able to control $\delta(e,Sx_1)^{N_1+\mu}$ in (10.32). Any $\partial/\partial s_j$-derivative may involve a loss of $1 + |\eta + \xi| \leq C\, (1 + \|\eta + \xi\|_{x_1})\, \delta(e,Sx_1)^2$ from the exponential, no loss from its action on $\psi(P^{-1}s)$, and a possible loss of $\delta(e,Sx_1)^2$ (Lemma 9.3) from its action on $\bar{\chi}_{-\mu,P_{-1}}(s, \xi_1 + \eta)$. Q.E.D.

THEOREM 10.35

Under the assumptions of Theorem 10.1, A *can be written in a unique way as* Op(f) *where* f *is a symbol of uniform type and weight* m. *Moreover, for all* N, *assuming that* $\varphi \in \mathcal{S}_\lambda(\Lambda)$ *for some* $\lambda > \lambda_0(N)$, *one may write*

$$|||\, f\, |||_{m,N} \leq C(\varphi, N, N_1, C_1)\, |||\, A\, |||_{m,\varphi,\tilde{N}} \tag{10.36}$$

for some $\tilde{N}$ *depending only on* (N_1,N).

PROOF What has to be done is to show that if, for all μ, a_μ is a symbol of weight m, then f, defined by the integral formula (7.5), is also a symbol of weight m. More precisely, we have to show that, given any N, there exists μ_0 such that, for $\mu > \mu_0$, the map $a_\mu \mapsto f$ is a sequentially continuous map from the space of symbols of weight m to the space of symbols of weight m up to the order of differentiability N. The proof (the last of its type in a series of five!) is much simpler than that of Theorem 7.10 since, now, μ is here not to create a problem, but to help to solve one. Let us recall that

$$f(y, \eta) = 2^{-n} \int_{T^*\Lambda} a_\mu(s, \xi)\, e^{2i\pi\langle \xi-\eta, s-S_y s\rangle}\, \delta(s, S_y s)^{-\mu}$$

$$\mathrm{Det}\left(\frac{\partial}{\partial y}(S_y s)\right) \det\left(\frac{\partial}{\partial s}(s - S_y s)\right) ds\, d\xi. \qquad (10.37)$$

Since

$$S_y s = S_y P_s^{1/2} e = P_s^{1/2}\, S_{P_s^{-1/2} y}\, e, \qquad (10.38)$$

one has

$$\mathrm{Det}\left(\frac{\partial}{\partial y}(S_y s)\right) = \mathrm{Det}\left(\frac{\partial}{\partial x}(S_x e)\right)(x = P_s^{-1/2} y)$$

$$= \mathrm{Det}(2\, M_{P_s^{-1/2} y}), \qquad (10.39)$$

a polynomial in $P_s^{-1/2} y$: also note that $P_s^{-1/2} e = Ss$. To estimate $f(y,\eta)$ we may, using covariance, assume that $y = e$. As usual, we shall write

$$|a_\mu(s, \xi)| \leq C_1 |||a_\mu|||_{m;0}\, m(e, \eta)\, \delta(e, s)^{N_1} (1 + |\xi - \eta|^2)^{N_1}. \qquad (10.40)$$

As $\delta(s,Ss) \geq C^{-1} \delta(e,s)^2$ and μ can be taken arbitrarily large, all we have to do is to save powers of $(1 + |\xi - \eta|^2)^{-1}$ with the help of integrations by parts based on the operators $\partial/\partial\theta_j$ with $\theta = s - Ss$: then of course $\mathrm{Det}(\partial/\partial s\, (s - Ss))ds$ is just $d\theta$. To see the effect of $(\partial/\partial\theta)^\alpha$ on $a_\mu(s,\xi)$, we note that the matrix $(\partial s_j/\partial\theta_k)$ is the inverse of the matrix $I + P_s^{-1}$ by [F1; Proposition II.3.3]. As P_s^{-1} is positive definite, $(I + P_s^{-1})^{-1}$ is a contraction so that the coefficients $\partial s_j/\partial\theta_k$ are uniformly bounded. The coefficients of $(\partial/\partial\theta)^\alpha$ in terms of the $(\partial/\partial s)^\beta$ with $|\beta| \leq |\alpha|$ are bounded by powers (depending on $|\alpha|$ only) of $\delta(e,s)$ by what has just been said together with Lemma 5.5. As $(\partial/\partial s)^\beta$ can be applied to a_μ (s,ξ) at a loss of $\delta(e,s)^{2|\beta|}$ at worst, and as the entries of Ss and their derivatives can be controlled by powers of $\delta(e,s)$ (Lemma 5.5 again), we get the estimate

$$|||f|||_{m,0} \leq C(C_1, N_1, \mu)\, |||a_\mu|||_{m,N}, \qquad (10.41)$$

provided that μ and N are large enough, in a way depending on N_1 only. Finally, it is trivial to control the factors that would come from a $\partial/\partial\eta$-differentiation in (10.37) (thanks to Lemma 4.27), and Lemmas 5.2 and 5.10, together with (10.38), take care of the $\partial/\partial y$-derivatives. Q.E.D.

COROLLARY 10.42

Let $A = Op(f)$ *and* $B = Op(g)$ *with* f *(resp.* g*) a symbol of uniform type and weight* m_1 *(resp.* m_2*). Then* $AB = Op(f \# g)$ *where* $f \# g$ *is of uniform type and weight* $m_1 m_2$*. More precisely, assume that the weights* m_1 *and* m_2 *are of type* C_1, N_1*. Then, given any* N*, there exists* $\tilde{N} = \tilde{N}(N, N_1)$ *such that*

$$|||f \# g|||_{m_1 m_2, N} \leq C(N, C_1, N_1) |||f|||_{m_1, \tilde{N}} |||g|||_{m_2, \tilde{N}}. \qquad (10.43)$$

PROOF First observe that the composition AB is well-defined, by Corollary 9.36, as an operator from $\mathcal{S}(\Lambda)$ to $\mathcal{S}(\Lambda)$. In view of Theorem 9.19 and its converse Theorem 10.35, (10.43) is equivalent, with some other $\tilde{N} = \tilde{N}(N, N_1)$, to a set of estimates (for which we may assume that $\varphi \in \mathcal{S}(\Lambda)$ in order not to be burdened with λ)

$$|||AB|||_{m_1 m_2, \varphi, N} \leq C(N, C_1, N_1, \varphi) |||A|||_{m_1, \varphi, \tilde{N}} |||B|||_{m_2, \varphi, \tilde{N}}. \qquad (10.44)$$

Since [compare (0.35)], using Lemma 10.4,

$$(U(\gamma_1)\varphi \mid ABU(\gamma)\varphi) = c(\varphi)^{-1} \int (U(\gamma_1)\varphi \mid AU(\gamma_2)\varphi)\, (U(\gamma_2)\varphi \mid BU(\gamma)\varphi)\, d\gamma_2, \qquad (10.45)$$

this follows from the definition, in Theorem 10.1, of $|||A|||_{m_1, \varphi, \tilde{N}}$, from (10.12), from the triangle inequality relative to d and, finally, from the second part of Theorem 5.21. Q.E.D.

11

Asymptotic Expansions

Even though we have used mostly the active Fuchs symbol up to now, it is in no way "better" than the passive symbol, or than a whole class of related symbols.

PROPOSITION 11.1
Let F *be a smooth function on* $\mathbb{R}^n$*, satisfying estimates*

$$\left|\left(\frac{\partial}{\partial\theta}\right)^{\alpha} F(\theta)\right| \leq C(\alpha)\,(1 + |\theta|^2)^{N(\alpha)}; \tag{11.2}$$

assume, moreover, that F *is K-invariant so that* $F(P^{-1}\theta)$ *is well defined, for* $\theta \in \mathbb{R}^n$ *and* $P \in GL(\Lambda)$*, as a function of* (Pe,θ) *only. Then the map* $f \mapsto h$ *such that*

$$(\mathcal{F}_2^{-1}h)(Pe, \theta) = F(P^{-1}\theta)\,(\mathcal{F}_2^{-1}f)(Pe, \theta) \tag{11.3}$$

extends, for every weight function m*, as a sequentially continuous map from the space of symbols of uniform type and weight* m *to itself. More precisely, let* m *be of type* (C_1,N_1) *and let* N *be any nonnegative integer: then, with some* $\tilde{N} = \tilde{N}(F,N,N_1)$ *and* $C = C(F,N,N_1,C_1)$*, one has*

$$|||h|||_{m,N} \leq C\,|||f|||_{m,\tilde{N}}. \tag{11.4}$$

PROOF As, for $(P_1,b_1) \in G$,

$$\mathcal{F}_2^{-1}(f \circ [P_1, b_1])(y, \theta) = |\det P_1|\, e^{-2i\pi\langle b_1, P_1\theta\rangle}(\mathcal{F}_2^{-1}f)(P_1 y, P_1\theta), \tag{11.5}$$

it is immediate that the map $f \mapsto h$ commutes with the action of G so that we may as well study h and its derivatives only at (y,ξ) = (e,ξ) (we might even take $\xi = 0$, but we do not think this would make matters more clear). After standard integrations by parts, we get

$$h(e, \xi) = \int_{T^*\Lambda} (1 + |\theta|^2)^{-\ell} e^{2i\pi\langle\theta,\eta-\xi\rangle} \left[1 - (2\pi)^{-2} \sum \frac{\partial^2}{\partial\theta_j^2}\right]^k F(\theta)$$

$$\left[1 - (2\pi)^{-2} \sum \frac{\partial^2}{\partial\eta_j^2}\right]^{\ell} ((1 + |\eta - \xi|^2)^{-k} f(e, \eta))\, d\eta\, d\theta. \tag{11.6}$$

Since

$$\left|\left(\frac{\partial}{\partial\eta}\right)^{\alpha} f(e, \eta)\right| \leq |||f|||_{m,|\alpha|}\, m(e, \eta)$$

$$\leq C_1 |||f|||_{m,|\alpha|}\, m(e, \xi)\, (1 + |\eta - \xi|^2)^{N_1}, \tag{11.7}$$

it is clear that, choosing k first so that $k - N_1 > n/2$, then ℓ large enough, we get

$$|h(e, \xi)| \leq C(F, N_1, C_1)\, m(e, \xi) |||f|||_{m,N} \tag{11.8}$$

for some N depending on (F,N_1). As we are dealing with a convolution operator, the examination of derivatives offers no difficulty whatsoever.

Q.E.D.

PROPOSITION 11.9
The active Fuchs symbol of an operator is of uniform type and weight m *if and only if its passive symbol has the same properties.*

PROOF By Proposition 3.28, we have to show that F and F^{-1} satisfy the hypotheses of Proposition 11.1 if F is defined by

$$F(\theta) = \omega(t)^{-2}\left[\operatorname{Det}\left(\frac{\partial}{\partial x}(S_x t)\right)(x = e)\, \operatorname{Det}\left(\frac{\partial}{\partial t}(t - St)\right)\right]^{-1} \tag{11.10}$$

with $St - t = \theta$. Recall from Lemma 4.27 that

$$C_0^{-1}\, (1 + |\theta|^2)^{1/2} \leq \delta(e, t)^2 \leq C_0\, (1 + |\theta|^2)^{N_0} \tag{11.11}$$

for some N_0 depending only on Λ. To bound the derivatives of F by powers of $\delta(e,t)$, the argument is the same as the one in the proof of

Theorem 10.35 that led to the fact that the coefficients $\partial t_k/\partial\theta_j$ are bounded. Lemma 5.5 is used also, again; details are left to the reader. The K-invariance of F is obvious. Q.E.D.

PROPOSITION 11.12
Consider the symbolic calculus in which the species of symbol to be used is linked to the active symbol f *of an operator* A *by the map* $f \mapsto h$ *introduced in Proposition 11.1. It is always* G*-covariant; it is* Γ*-covariant (*Γ *being the group introduced in Proposition 3.3) if and only if the function* F *is even.*

PROOF We have already noticed that the map $f \mapsto h$ commutes with the action of G. If $\tilde{S}(y,\eta) = (Sy, -P_y\eta)$, it is immediate that

$$(\mathcal{F}_2^{-1}(f \circ \tilde{S}))(y, \theta) = \omega(y)^{-2}(\mathcal{F}_2^{-1}h)(Sy, -P_y^{-1}\theta) \tag{11.13}$$

so that one goes from this function to $(\mathcal{F}_2^{-1}(h \circ \tilde{S}))(y,\theta)$ by multiplying by $F(-P_y^{-1/2}\theta)$; on the other hand, if h_1 is the function that corresponds to $f \circ \tilde{S}$ under the map $f \mapsto h$, we get $(\mathcal{F}_2^{-1}h_1)(y,\theta)$ by multiplying the function in (11.13) by $F(P_y^{-1/2}\theta)$. Q.E.D.

In Definition 6.28 we introduced symbols of classical type and order k: the difference with symbols of uniform type is that it is not only harmless, but also even beneficial in some specified sense, to differentiate a symbol $f(y,\eta)$ with respect to the η-variables. As is well-known from classical pseudodifferential operator theory on $\mathbb{R}^n$, this permits one to replace complicated operations on symbols, like $f \mapsto h$, by asymptotic expansions almost as good in a sense.

PROPOSITION 11.14
Let $f \in S^k$*, the space of symbols of classical type and order* k*. Let* F *satisfy the hypotheses of Proposition 11.1, and let* $F(\theta) \sim \Sigma\, a_\alpha \theta^\alpha$ *be the Taylor expansion of* F *at the origin. Then the function* h *associated to* f *by Proposition 11.1 also belongs to* S^k*; moreover, for every integer* $\ell \geq 1$*, the symbol defined as*

$$h(y, \eta) - \sum_{|\alpha|<\ell} (-2i\pi)^{-|\alpha|}\, a_\alpha(y) \left(P_y^{-1/2} \frac{\partial}{\partial\eta}\right)^{\alpha} f(y, \eta) \tag{11.15}$$

if $y = Pe$, $P \in GL(\Lambda)$, *belongs to* $S^{k-\ell}$.

PROOF It is identical to that of Proposition 14.2 of [U6]: let us just recall the principle of the proof to see why it extends without change. First, the remainder is zero for large ℓ in the case when h is a polynomial as a function of η for fixed y: then, in all cases, the sum of all terms for fixed $|\alpha|$ belongs to $S^{k-|\alpha|}$ so that it suffices (since one could always push expansions further as far as needed) to prove that the remainder is of uniform type and weight $(1 + |\eta|_y^2)^{1/2(k-\ell)}$. For given η°, to estimate the remainder at (y,η°), one proceeds as follows. Let

$$f = T_{\eta^\circ}^{\ell-1}(f) + R_{\eta^\circ}^{\ell}(f), \tag{11.16}$$

where the two terms are the regular part and the remainder of the Taylor expansion of f to the order $\ell - 1$, with respect to the η-variables, at η°. Then, the operator given by (11.3), when applied to the first term, will yield a symbol that coincides at $\eta = \eta^\circ$ with the sum that is considered in (11.15). Finally, a careful but elementary examination of the remainder $(R_{\eta^\circ}^{\ell}(f))(y,\eta)$ shows that it is, in a uniform way, a symbol of uniform type and weight

$$m(y, \eta) = (1 + |\eta|_y^2)^{(k-\ell)/2}(1 + |\eta - \eta^\circ|_y^2)^{(\ell+|k-\ell|)/2}. \tag{11.17}$$

Then it suffices to apply to the remainder the precise estimate given in Proposition 11.1, and to note that the nuisance factor on the right-hand side of (11.17) reduces to 1 at η°. Details may be found in [U6]. Q.E.D.

It should be noted that it is precisely with a view towards such kinds of proofs that we have been careful, in Sections 7 to 10, to give estimates uniform with respect to all weights of type (C_1,N_1). Indeed, here, we had to apply the estimate in Proposition 11.1 with the η°-dependent weight in (11.17). In the proof of the composition formula, we shall need, in the same way, the weight-independent estimate of Corollary 10.42.

PROPOSITION 11.18

Let f *be a differential symbol, i.e., a polynomial* $f(y,\eta) = \Sigma\, a_\alpha(y)\eta^\alpha$ *with respect to the* η*-variables. It is assumed that the coefficients* a_α *are smooth and have derivatives bounded by powers of* $\delta(e,y)$ *in such a way that* f *belongs to a certain class of symbols of uniform type, for a suitable weight. The operator* Op(f) *is a differential operator, characterized by the formula*

$$(\mathrm{Op}(f)u)(s) = \sum \left(\frac{1}{2i\pi}\frac{\partial}{\partial t}\right)^{\alpha}\left[a_\alpha(\mathrm{mid}(s,t))2^n\,\mathrm{Det}\left(\frac{\partial}{\partial t}\,\mathrm{mid}(s,t)\right)u(t)\right](t = s).$$

PROOF Let us compute the standard symbol a of Op(f) defined by (1.32). In view of Theorem 7.10, we shall dispense with the easy arguments that make (1.32) valid, in the distribution sense, in the case under consideration. As

$$\eta^\alpha e^{2i\pi\langle\eta-\xi,s-t\rangle} = \left(-\frac{1}{2i\pi}\frac{\partial}{\partial t} + \xi\right)^\alpha e^{2i\pi\langle\eta-\xi,s-t\rangle}, \tag{11.19}$$

an integration by parts in (1.32) permits one to write

$$a(s, \xi) = 2^n \sum \binom{\alpha}{\beta}\xi^\beta$$

$$\left(\frac{1}{2i\pi}\frac{\partial}{\partial t}\right)^{\alpha-\beta}\left\{a_\alpha(\mathrm{mid}(s, t))\mathrm{Det}\left(\frac{\partial}{\partial t}(\mathrm{mid}(s, t))\right)\right\}(t = s) \tag{11.20}$$

so that, from (1.30), we get

$$(\mathrm{Op}(f)u)(s) = 2^n \sum \binom{\alpha}{\beta}\left(\frac{1}{2i\pi}\frac{\partial}{\partial t}\right)^{\alpha-\beta}$$

$$\left\{a_\alpha(\mathrm{mid}(s, t))\mathrm{Det}\left(\frac{\partial}{\partial t}(\mathrm{mid}(s, t))\right)\right\}(t = s)\left(\frac{1}{2i\pi}\frac{\partial}{\partial s}\right)^\beta u, \tag{11.21}$$

which leads right away to Proposition 11.18. Q.E.D.

With y = mid(s,t), then s = Pe, $P \in GL(\Lambda)$, let us note that

$$\mathrm{Det}\left(\frac{\partial}{\partial t}(\mathrm{mid}(s, t))\right) = \mathrm{Det}\left(\frac{\partial}{\partial y}(S_y s)\right)^{-1} = \mathrm{Det}\left(\frac{\partial}{\partial y}(PS_{P^{-1}y}e)\right)^{-1}$$

$$= \left[\mathrm{Det}\left(\frac{\partial}{\partial x}(S_x e)\right)(x = P^{-1}y)\right]^{-1} = 2^{-n}(\det M_{P^{-1}y})^{-1} \tag{11.22}$$

since the derivative of the map $x \mapsto S_x e$ is $2M_x$ in the Jordan algebraic sense. In particular, when t = s, one has $P^{-1}y = e$, so that the operator with symbol f is the multiplication by f(y) in the case when f depends only on y. This is the reason why the Fuchs calculus is defined with a 2^n in front of the integral (1.7). First-order differential operators have the same symbol in all Γ-covariant symbolic calculi connected to the active symbolic calculus by means of Proposition 11.1. Indeed, as F is an even function, the sum in Proposition 11.14 reduces to its first term in the case under consideration. This is true, in particular, for the infinitesimal operators of the representation U introduced in (3.25). These are the infinitesimal generators, in the sense of Stone's theorem, of the unitary groups that are the images, under U, of one-parameter subgroups of G.

PROPOSITION 11.23
The symbol of the operator of multiplication by $\langle b,t\rangle$ *is* $f(y,\eta) = \langle b,y\rangle$. *If* $A = (a_{jk}) \in \mathfrak{gl}(\Lambda)$, *the Lie algebra of* $GL(\Lambda)$, *the symbol of the operator*

$$(2i\pi)^{-1}At\,\frac{\partial}{\partial t} = (2i\pi)^{-1}\sum a_{jk}t_k\,\frac{\partial}{\partial t_j}$$

is $f(y,\eta) = \langle Ay,\eta\rangle$. *In any of these cases, given another symbol* g *(in some class of uniform type, say) one has*

$$[Op(f), Op(g)] = (2i\pi)^{-1}Op(\{f, g\}), \tag{11.24}$$

where {f,g} *denotes the Poisson bracket of* f *and* g *with respect to the symplectic structure on* $\Lambda \times \mathbb{R}^n$ *induced by the embedding of that space in* $\mathbb{R}^n \times \mathbb{R}^n$.

PROOF The first point has been proved already. From the same computation as the one in (11.22), only looking at the derivatives involved themselves (and not only at their jacobians), it follows that

$$\frac{\partial}{\partial t_j}(\text{mid}(s, t))_k = \frac{1}{2}\,\delta_{jk} \qquad \text{at } t = s. \tag{11.25}$$

If $f(y,\eta) = \langle Ay,\eta\rangle$, Proposition 11.18 yields

$$\begin{aligned} 2i\pi(Op(f)u)(s) &= \left(As\,\frac{\partial}{\partial s}\right)u + \frac{1}{2}\,(\text{Tr } A)\,u(s) \\ &\quad + 2^n\left\langle As, \frac{\partial}{\partial t}\left(\text{Det}\left(\frac{\partial}{\partial t}\,(\text{mid}(s, t))\right)\right)(t = s)\right\rangle u(s). \end{aligned} \tag{11.26}$$

With $y = \text{mid}(s,t)$, we have to compute

$$\begin{aligned} &2^n\,At\,\frac{\partial}{\partial t}\left(\text{Det}\left(\frac{\partial}{\partial t}\,(\text{mid}(s, t))\right)\right)(t = s) \\ &= 2^{n-1}Ay\,\frac{\partial}{\partial y}\left[\text{Det}\,\frac{\partial}{\partial y}\,(S_y s)\right]^{-1}(y = s) \\ &= -2^{-n-1}Ay\,\frac{\partial}{\partial y}\,\text{Det}\left(\frac{\partial}{\partial y}\,(S_y s)\right)(y = s) \end{aligned}$$

$$= -\frac{1}{2} Ay \frac{\partial}{\partial y} (\det M_{P^{-1}y})(P^{-1}y = e) \tag{11.27}$$

where, in the last line, we have set s = Pe and used (11.22) again. With $x = P^{-1}y$, one has $Ay\, \partial/\partial y = (P^{-1} APx)\, \partial/\partial x$, and since $\mathrm{Tr}\, A = \mathrm{Tr}(P^{-1} AP)$, we see from (11.26) that it only remains to be shown that

$$Ax \frac{\partial}{\partial x} (\det M_x)(x = e) = \mathrm{Tr}\, A \tag{11.28}$$

for every $A \in \mathfrak{gl}(\Lambda)$. Now, the left-hand side may be written as

$$\frac{d}{d\epsilon} (\det M_{e^{\epsilon A}e})(\epsilon = 0) = \frac{d}{d\epsilon} (\det(I + \epsilon M_{Ae}))(\epsilon = 0) = \mathrm{Tr}\, M_{Ae}. \tag{11.29}$$

Here $e^{\epsilon A}e$ is the exponential of the vector field ϵA, applied to the unit element $e \in \Lambda$, and $M_{e^{\epsilon A}e}$ is the corresponding multiplication operator. Thus we just have to show that $\mathrm{Tr}\, M_{Ae} = \mathrm{Tr}\, A$. In the case when $A = -A'$, $e^{\epsilon A}$ belongs to K so that $Ae = d/d\epsilon\, (e^{\epsilon A}e)(\epsilon = 0) = 0$. We may thus assume that $A = A'$; then $A = M_x$ for some $x \in \Lambda$ and $M_{Ae} = A$ as operators. We have thus proved that the operator with symbol $\langle Ay,\eta\rangle$ is $(2i\pi)^{-1}$ At $\partial/\partial t$ whenever $A \in \mathfrak{gl}(\Lambda)$. A trivial covariance argument permits to prove the end of Proposition 11.23. Assume, for instance, that $f(y,\eta) = \langle Ay,\eta\rangle$ as above. Then

$$(\mathrm{Op}(f)u)(s) = (2i\pi)^{-1} \frac{d}{d\epsilon} (u(e^{\epsilon A}s))(\epsilon = 0), \tag{11.30}$$

in other words

$$\mathrm{Op}(f) = -(2i\pi)^{-1} \frac{d}{d\epsilon} U(e^{\epsilon A}, 0)(\epsilon = 0). \tag{11.31}$$

If g is another symbol, one thus has

$$[\mathrm{Op}(f), \mathrm{Op}(g)] = -(2i\pi)^{-1} \frac{d}{d\epsilon} (U(e^{\epsilon A}, 0)\, \mathrm{Op}(g)\, U(e^{-\epsilon A}, 0))(\epsilon = 0) \tag{11.32}$$

and, by covariance,

$$U(e^{\epsilon A}, 0)\, \mathrm{Op}(g)\, U(e^{-\epsilon A}, 0) = \mathrm{Op}(h) \tag{11.33}$$

with

$$h(y, \eta) = g(e^{-\epsilon A}y, e^{\epsilon A'}\eta). \tag{11.34}$$

Hence the active symbol of [Op(f),Op(g)] is

$$(2i\pi)^{-1}\left(Ay\frac{\partial}{\partial y} - A'\eta\frac{\partial}{\partial \eta}\right)g = (2i\pi)^{-1}\sum\left(\frac{\partial f}{\partial \eta_j}\frac{\partial g}{\partial y_j} - \frac{\partial f}{\partial y_j}\frac{\partial g}{\partial \eta_j}\right), \tag{11.35}$$

which is just what has been asserted in Proposition 11.23. Q.E.D.

Let us remark that the same formula holds whenever Γ-covariant symbols of any type related to the active symbol under the map in Proposition 11.1 are used throughout: one does not have to change f, though.

PROPOSITION 11.36

There exists a unique family $(P_{\alpha,\beta}(y, \partial/\partial s, \partial/\partial t))$ *depending on the pair* (α,β) *of multi-indices, with the following properties:*

(i) *for every* α,β, $P_{\alpha,\beta}(y, \partial/\partial s, \partial/\partial t)$ *is a differential operator with constant coefficients on functions of* $(s,t) \in \mathbb{R}^n \times \mathbb{R}^n$. *Its order is at most* $|\alpha| + |\beta|$ *and its coefficients are* C^∞ *functions of* $y \in \Lambda$.

(ii) *for every pair* (f,g) *of differential symbols (cf. Proposition 11.18), the symbol* f # g, *which is the active symbol of* Op(f)Op(g), *is given by*

$$(f \# g)(y, \eta) = \sum\left[P_{\alpha,\beta}\left(y, \frac{\partial}{\partial s}, \frac{\partial}{\partial t}\right)(\partial_\eta^\alpha f(s, \eta)\partial_\eta^\beta g(t, \eta))\right](s = t = y). \tag{11.37}$$

Moreover, the term on the right-hand side that corresponds to $\alpha = \beta = 0$ *reduces to the product* $f(y,\eta)g(y,\eta)$; *the sum of terms with* $|\alpha| + |\beta| = 1$ *is* $(4i\pi)^{-1}\{f,g\}$, *where the Poisson bracket is the one associated with the standard symplectic structure on* $T^*\Lambda = \Lambda \times \mathbb{R}^n$.

PROOF If a is the standard symbol of Op(f), one may rewrite (11.20) as

$$a(y, \eta) = \sum\frac{1}{\beta!}\left(\frac{1}{2i\pi}\frac{\partial}{\partial t}\right)^\beta[\chi(y, t)\,\partial_\eta^\beta f(\mathrm{mid}(y, t), \eta)](t = y) \tag{11.38}$$

with

$$\chi(y, t) = 2^n \operatorname{Det}\left(\frac{\partial}{\partial t} \operatorname{mid}(y, t)\right). \tag{11.39}$$

In particular, if f, as a polynomial in η, is expanded as a sum of homogeneous terms as

$$f = f_k + f_{k-1} + \cdots + f_0, \tag{11.40}$$

and the same is done for a, then $a_k = f_k$, and

$$a_{k-1}(y, \eta) = f_{k-1}(y, \eta) + \frac{1}{2i\pi} \sum \varphi_j(y) \frac{\partial f_k}{\partial \eta_j} + \frac{1}{4i\pi} \sum \frac{\partial^2 f_k}{\partial y_j \partial \eta_j} \tag{11.41}$$

with $\varphi_j(y) = [\partial/\partial t_j\ (\chi(y,t))]\ (t = y)$. One has also used (11.25) to compute the coefficient of the last term. If one sets $a = f + Nf$, the operator N is nilpotent when considered as defined on the space of differential symbols with order less than any given integer. Thus (11.38) can be inverted as

$$f = \sum_{j \geq 0} (-1)^j N^j a.$$

Assertions (i) and (ii) in Proposition 11.36 are then a consequence of the formula (0.5) for the composition of standard symbols, together with the fact that, in each term on the right-hand side of (11.38), one must differentiate f with respect to η at least as many times as with respect to the first n variables. To find the terms with $|\alpha| + |\beta| \leq 1$ in the expansion (11.37) of f # g, one may start from (11.38) truncated as in (11.41) (with f substituted for f_k or f_{k-1} on the right-hand side, a substituted for a_{k-1} on the left-hand side) and use (0.5) with the right-hand side reduced to $ab + (4i\pi)^{-1}\{a,b\}$. The last part of Proposition 11.36 is then a consequence of the identity

$$\frac{1}{4i\pi} \sum \left(f \frac{\partial^2 g}{\partial y_j \partial \eta_j} + g \frac{\partial^2 f}{\partial y_j \partial \eta_j} \right) + \frac{1}{2i\pi} \sum \frac{\partial f}{\partial \eta_j} \frac{\partial g}{\partial y_j}$$
$$= \frac{1}{4i\pi} \sum \frac{\partial^2}{\partial y_j \partial \eta_j} (fg) + \frac{1}{4i\pi} \{f, g\}. \quad \text{Q.E.D.}$$

REMARKS 11.42

1) one has a similar conclusion for any Γ-covariant type of symbol related to the active symbol by Proposition 11.1 (with F even); 2) covariance under G shows that the polynomials Q_{jk} in the 4n variables σ, τ, ξ, ζ, with coefficients depending on y, defined as

$$Q_{jk}(y;\sigma,\tau;\xi,\zeta) = \sum_{|\alpha|=j,|\beta|=k} P_{\alpha,\beta}(y,\sigma,\tau)\xi^\alpha\zeta^\beta \tag{11.43}$$

are invariant under the transformations $(y;\sigma,\tau;\xi,\zeta) \mapsto (Py; P'^{-1}\sigma, P'^{-1}\tau; P\xi, P\zeta)$ with $P \in GL(\Lambda)$; 3) to find explicitly the $P_{\alpha,\beta}$'s may be too complicated for what it is worth; in the half-line case, the complete formula can be found in [U5], Theorem 7.2: one should be careful, however, that a symbol denoted as $f(\xi,y)$ there would be denoted here as $f(y,y\xi)$.

THEOREM 11.44
Let $f \in S^{k_1}$ *and* $g \in S^{k_2}$ *be two classical symbols. For every integer* $\ell \geq 1$, *let us define, with the notations of Proposition 11.36,*

$$h_\ell(y,\eta) = \sum_{|\alpha|+|\beta|<\ell} \left[P_{\alpha,\beta}\left(y,\frac{\partial}{\partial s},\frac{\partial}{\partial t}\right)(\partial_\eta^\alpha f(s,\eta)\partial_\eta^\beta g(t,\eta))\right](s=t=y).$$

Then the composition $f \# g$, *defined in Corollary 10.42, satisfies the property that, for every* ℓ, $f \# g - h_\ell$ *is a classical symbol of order* $k_1 + k_2 - \ell$.

PROOF For each integer $j \geq 0$, denote as $(f \# g)_j$ the sum of the terms with $|\alpha| + |\beta| = j$ in the asymptotic sum under consideration. It is of course well-defined for any two C^∞ functions f,g on $\Lambda \times \mathbb{R}^n$. Let $\gamma = [P,b]$ be any element in G, and set $\tilde{f} = f \circ [\gamma]$, $\tilde{g} = g \circ [\gamma]$. The uniqueness of the operators $P_{\alpha,\beta}$ $(y, \partial/\partial s, \partial/\partial t)$ in Proposition 11.36 implies that $(\tilde{f} \# \tilde{g})_j = (f \# g)_j \circ [\gamma]$ since this is true whenever f and g are differential symbols: it then follows from (6.22) that $(f \# g)_j \in S^{k_1+k_2-j}$ if $f \in S^{k_1}$ and $g \in S^{k_2}$. As a consequence, it suffices (since one can push the expansion as far as needed) to show that, for every ℓ, $f \# g - h_\ell$ is a symbol of uniform type and weight $(1 + |\eta|_y^2)^{(k_1+k_2-\ell)/2}$. In view of the identity $\tilde{f} \# \tilde{g} = (f \# g) \circ [\gamma]$, a consequence of the covariance under G of the Fuchs calculus, it suffices to estimate the derivatives of h_ℓ at $y = e$, where one does not have to worry about the size of the coefficients of the operators $P_{\alpha,\beta}(y, \partial/\partial s, \partial/\partial t)$.

The proof then parallels that of Proposition 0.45, starting from the decomposition of f (or g) as a sum of two terms arising from the Taylor expansion (11.16). Corollary 10.42 yields, at $\eta = \eta^\circ$, the required estimate for the three remainder terms that occur in the decomposition of $f \# g$ as a sum of four terms. It is crucial that these estimates do not depend on the weight, only on the pair (C_1, N_1), which occurs in Definition 6.5. The main term in $(f \# g)(y,\eta^\circ)$ comes from the evaluation at (y,η°) of the composition $T_{\eta^\circ}^{\ell-1}(f) \# T_{\eta^\circ}^{\ell-1}(g)$. It coincides, as a consequence of Proposition 11.36, with the right-hand side of the formula in Theorem 11.44, plus a bunch of extra terms that can be discarded since they are, up to the order of differentiability zero, of order $k_1 + k_2 - \ell$. Since the decompositions (11.16) only depend on η°, the preceding discussion also takes care of the $\partial/\partial y$-derivatives of $f \# g - h_\ell$. Finally, letting t_k denote the operator of multiplication of $u = u(t)$ by t_k, Proposition 11.23 shows that $(2i\pi)^{-1}\partial f/\partial\eta_k$ is the symbol of $-[t_k, \mathrm{Op}(f)]$. It follows that

$$\frac{\partial}{\partial\eta_k}(f \# g) = \frac{\partial f}{\partial\eta_k} \# g + f \# \frac{\partial g}{\partial\eta_k}. \tag{11.45}$$

Since, obviously, one has for every j

$$\frac{\partial}{\partial\eta_k}(f \# g)_j = \left(\frac{\partial f}{\partial\eta_k} \# g\right)_j + \left(f \# \frac{\partial g}{\partial\eta_k}\right)_j, \tag{11.46}$$

the proof of Theorem 11.44 is complete. Q.E.D.

12

A Beals-type Characterization of Operators of Classical Type

In [B2], R. Beals proved a useful characterization of classes of pseudodifferential operators on $\mathbb{R}^n$ by means of the continuity properties of their iterated brackets with operators taken from a list characteristic of the class under study. These would be just the operators x_j or $\partial/\partial x_j$ in the simplest cases. We show here that the same can be done in the Fuchs calculus for operators of classical type. We need a very small amount of weighted Sobolev spaces theory, for which we have not strived for generality.

DEFINITION 12.1 A *multiplicator* μ shall be any weight function $\mu(y)$ depending only on y, satisfying the property that μ is a symbol of uniform type and of weight μ. Given a multiplicator μ and a real number k, we denote as $S^k(\mu)$ the space of smooth symbols f satisfying, for every pair (p,q) of nonnegative integers, the inequality

$$\|f\|_{p,q;\,y,\eta} \leq C(p,q)\ \mu(y)\ (1 + |\eta|_y^2)^{(k-q)/2}$$

for some constant C(p,q). In particular, $S^k(1)$ coincides with the space of symbols S^k introduced in Definition 6.28. A symbol f shall be called *elliptic* in $S^k(\mu)$ if $f \in S^k(\mu)$, and there exists $g \in S^{-k}(\mu^{-1})$ such that $fg - 1$ belongs to $S^{-1}(1)$.

PROPOSITION 12.2

A class $S^k(\mu)$ *is invariant under any map of the kind* $f \mapsto h$ *introduced in Proposition 11.1 and in particular under the map connecting passive to active symbols. If* $f \in S^k(\mu)$ *and* $g \in S^j(\nu)$ *then the composition* $f \# g$ *belongs to* $S^{k+j}(\mu\nu)$ and $f \# g - fg$ *belongs to* $S^{k+j-1}(\mu\nu)$. *If* f (*resp.* g) *is elliptic in* $S^k(\mu)$ (*resp.* $S^j(\nu)$), *then* $f \# g$ *is elliptic in* $S^{k+j}(\mu\nu)$.

PROOF The first part is a consequence of Proposition 11.1 and of the fact that the map $f \mapsto h$ commutes with differentiation operators with constant coefficients in the fibers of $T^*(\Lambda)$. The second part may be proved in exactly the same way as Theorem 11.44. The last part then follows.

Q.E.D.

PROPOSITION AND DEFINITION 12.3

Let k *be a nonnegative real number. Let* f *be elliptic in* $S^k(1)$. *The space of* $u \in L^2(\Lambda)$ *such that* $\mathrm{Op}(f)u \in L^2(\Lambda)$ *does not depend on the choice of* f—*we denote it as* H^k. *Moreover, if we let*

$$\|u\|_k = \|u\| + \|\mathrm{Op}(f)u\| ,$$

the topology associated with this norm on H^k *does not depend on* f *either. We then define* H^{-k} *as the topological dual of* H^k, *more precisely as the space of all distributions in* Λ *that extend as continuous linear forms on* H^k *and provide it with the norm dual to that of* H^k.

PROOF Let f be an elliptic symbol in $S^k(1)$ and let $g \in S^{-k}(1)$ be such that $gf - 1$ belongs to $S^{-1}(1)$: then, as a consequence of Proposition 12.2, $g \# f - 1$ belongs to $S^{-1}(1)$ too. Following the induction procedure completely standard in the construction of parametrices, one can then build a sequence $(g_p)_{p \geq 0}$, with $g_p \in S^{-k-p}(1)$, such that

$$r_{p+1} := (g_0 + \ldots + g_p) \# f - 1$$

belongs to $S^{-1-p}(1)$ for all p: it suffices to set $g_0 = g$ and, for $p \geq 0$, $g_{p+1} = -r_{p+1}g$; then $r_{p+2} = r_{p+1} + g_{p+1} \# f$ coincides (as a consequence of Proposition 12.2) with $r_{p+1} + g_{p+1}\, f = r_{p+1}\,(1 - g\, f)$ up to an error term in $S^{-2-p}(1)$, so we are done. Choosing now p with $p + 1 \geq k$, set $f_{-1} = g_0 + \ldots + g_p \in S^{-k}(1)$ and $r = f_{-1} \# f - 1$, a symbol that lies in $S^{-k}(1)$ too. Then, if g is any symbol in $S^k(1)$, one can write

$$\mathrm{Op}(g)u = \mathrm{Op}(g \# f_{-1})\mathrm{Op}(f)u - \mathrm{Op}(g \# r)u,$$

a function in $L^2(\Lambda)$ since, by Proposition 12.2, $g \# f_{-1}$ and $g \# r$ belong to $S^0(1)$. The same identity shows the continuity of $\mathrm{Op}(g)u \in L^2(\Lambda)$ as a function of u in H^k, which concludes the proof of the proposition. Observe also that if $u \in H^k$ is the limit, in $L^2(\Lambda)$, of a sequence (w_j) in $\mathcal{S}(\Lambda)$, and if, moreover, $\mathrm{Op}(f)u$ is the limit in $L^2(\Lambda)$ of a sequence (v_j) in $\mathcal{S}(\Lambda)$, then u is the limit, in the space H^k, of the sequence

$$\mathrm{Op}(f_{-1})v_j - \mathrm{Op}(r)w_j.$$

This shows that $\mathcal{S}(\Lambda)$ is dense in H^k and makes the definition of H^{-k} meaningful. Q.E.D.

PROPOSITION 12.4

Let Δ be the Laplace operator on Λ associated with its Riemannian structure (not to be confused with the function on Λ by the same name!), and let k be a nonnegative integer. The topological vector space H^{2k} coincides with the space of all $u \in L^2(\Lambda)$ such that $\Delta^k u \in L^2(\Lambda)$, this latter space being provided with the norm $u \mapsto \|u\| + \|\Delta^k u\|$. Given $k_1, k_2 \in \mathbb{R}$ and $\theta \in [0,1]$, the Sobolev space $H^{(1-\theta)k_1+\theta k_2}$ coincides, as a topological vector space, with the intermediary space $[H^{k_1}, H^{k_2}]_\theta$ defined by complex interpolation theory [C2].

PROOF By Proposition 11.18, the active symbol f of Δ is, with respect to η, a polynomial of order 2 whose coefficients are smooth functions of y. The covariance property shows that the identity $f(Py, P'^{-1}\eta) = f(y,\eta)$ holds for any $P \in GL(\Lambda)$, and this implies that $f \in S^2(1)$. Still by Proposition 11.18, the principal part of f (i.e., its homogeneous second-order part as a polynomial in η) is the same as that of the standard symbol of Δ, and this is clearly (for any Riemannian structure on an open subset of $\mathbb{R}^n$) just $|\eta|_y^2$ up to normalization. As a consequence, f is elliptic in $S^2(1)$, and the first part of Proposition 12.4 follows. The proof of the last part is standard, based as it is on Hadamard's three-line theorem, since the function $f_z(y,\eta) = (1 + |\eta|_y^2)^{z/2}$ depends on z in a holomorphic way, say in a neighborhood of the strip $k_1 \leq \mathrm{Re}\, z \leq k_2$. However, when $\mathrm{Re}\, z = k_1$, f_z does not remain in a bounded subset of $S^{k_1}(1)$ so we consider instead of f_z the symbol $g_z = (z - k_1 + 1)^{-N} f_z$ for some large integer N. Indeed, to give a bound for the operator-norm of $\mathrm{Op}(f_{-k_1})\mathrm{Op}(f_z)$ we only need to consider finitely many derivatives of f_z, which yields a bound of the type $C(1 + |\mathrm{Im}\, z|)^N$. This concludes the proof of Proposition 12.4. Q.E.D.

PROPOSITION 12.5

For any $t \in \Lambda$, choose a basis $\{e_1, \ldots, e_n\}$ of $\mathbb{R}^n$ orthonormal with respect to $\| \ \|_t$; if $e_p = (\alpha_1, \ldots, \alpha_n)$, identify e_p with the constant vector field $\Sigma\, \alpha_j\, \partial/\partial t_j$. For every $k \in \mathbb{N}$ and $u \in C^\infty(\Lambda)$, define the function $N_k(u)$ on Λ through

$$(N_k^2(u))(t) = \sum_{p_1,\ldots,p_k=1}^{n} |(e_{p_1} \cdots e_{p_k} u)(t)|^2$$

where the right-hand side does not depend on the choice of the basis $\{e_j\}$. Then, on $C^\infty(\Lambda) \cap H^k$, the norms $\| \ \|_k$ and $u \mapsto \|u\| + \|N_k(u)\|$ are equivalent.

PROOF For each p, let e_p^* ($= -e_p +$ lower-order terms) be the adjoint of e_p on $L^2(\Lambda)$. Then $\|N_k(u)\|^2 = (u| \, Du)$ with

$$D = \Sigma \, e_{pk}^* \ldots e_{p_1}^* \, e_{p_1} \ldots e_{p_k}. \tag{12.6}$$

As D has smooth coefficients and is G-invariant, (6.22) shows that it has a symbol f in $S^{2k}(1)$. Moreover, the top-order part of f is

$$(4\pi^2)^k \sum (\langle e_{p_1}, \eta\rangle \cdots \langle e_{p_k}, \eta\rangle)^2 = (4\pi^2 |\eta|_y^2)^k, \tag{12.7}$$

which implies that f is elliptic in $S^{2k}(1)$. Using the symbolic calculus, it is now a routine job, starting from the symbol $(1 + 4\pi^2|\eta|_y^2)^{k/2}$ and using an induction procedure, to construct $g \in S^k(1)$ real-valued such that $g \# g - f \in S^0(1)$. Moreover, g is elliptic in $S^k(1)$. Then

$$|\|N_k(u)\|^2 - \|Op(g)u\|^2| \leq C_k \|u\|^2 \tag{12.8}$$

for some constant C_k, which implies Proposition 12.5 in view of Proposition 12.3. Q.E.D.

DEFINITION 12.9 Let k be a real number, and let μ be a multiplicator. We define $H^k(\mu)$ as the space of $u \in \mathcal{S}'(\Lambda)$ such that $\mu u \in H^k$ and set

$$\|u\|_{k;\mu} = \|\mu u\|_k \,,$$

which defines $H^k(\mu)$ as a topological vector space in a canonical way.

PROPOSITION 12.10
Let $u \in H^k(\mu)$ *and let* $f \in S^j(\nu)$: *then* $Op(f)u \in H^{k-j}(\mu\nu^{-1})$. *Conversely, if* $u \in H^{-N}(\mu)$ *for some* N, f *is elliptic in* $S^j(\nu)$ and $Op(f)u \in H^{k-j}(\mu\nu^{-1})$, *then* $u \in H^k(\mu)$. *Finally,* $H^k(\mu)$ *and* $H^{-k}(\mu^{-1})$ *are dual to each other as topological vector spaces.*

PROOF First observe that, if $u \in H^k$ and $f \in S^k(1)$, then $Op(f)u \in L^2(\Lambda)$. This follows from the proof of Proposition 12.3 in a direct way if $k \geq 0$; if $k < 0$, it suffices to use the definition of H^k as the space of all $u \in \mathcal{S}'(\Lambda)$ such that $|(v|u)| \leq C \|v\|_{-k}$ for some constant $C > 0$ and all $v \in \mathcal{S}(\Lambda)$, together with the fact that $Op(\bar{f})$ sends $L^2(\Lambda)$ into H^{-k}. One can then extend Proposition 12.3 as follows: if $u \in H^{-N}$ for some N and $Op(f)u \in L^2(\Lambda)$ for some f elliptic in $S^k(1)$ then $u \in H^k$. Indeed, in the case when $k \geq 0$, it suffices to push the construction already used in the proof of Proposition 12.3 further, ending up with $f_{-1} \in S^{-k}(1)$ such that $r = f_{-1} \# f - 1 \in S^{-N}(1)$.

This shows that $u = -\mathrm{Op}(r)u + \mathrm{Op}(f_{-1})\mathrm{Op}(f)u \in L^2(\Lambda)$, which is just what was missing. In the case when $k < 0$, it suffices to write, for any $v \in \mathcal{S}(\Lambda)$,

$$(v|u) = -(\mathrm{Op}(\bar{r})v|u) + (\mathrm{Op}\,\bar{f}_{-1})v \mid \mathrm{Op}(f)u),$$

so that, for some C depending only on u,

$$|(v|u)| \leq C\,[\|\mathrm{Op}(\bar{r})v\,\|_N + \|\mathrm{Op}(\bar{f}_{-1})v\|_0];$$

this is less than $C\,(\|v\|_0 + \|v\|_{-k})$ in view of what was seen in the first part of the present proof. Proposition 12.10 is an immediate consequence of the results proved so far in the case when $\mu = \nu = 1$. However, inserting weights (which can be inverted) here and there does not create any difficulty, in view of Proposition 12.2. Q.E.D.

The following lemma will be needed later.

LEMMA 12.11
Let B *be a bounded subset of* $C_0^\infty(\Lambda)$ *so that, in particular, the support of* $\varphi \in B$ *is contained in a fixed compact subset of* Λ. *For any* $\gamma = (P,b) \in G$ *and* $\varphi \in B$, *define* $\varphi_\gamma = U(\gamma)\varphi$, i.e.,

$$\varphi_\gamma(t) = \varphi(P^{-1}t)\, e^{2i\pi\langle b,t\rangle}.$$

Then, for any real number k *and multiplicator* μ *satisfying* (6.7) *for some pair* (C_1, N_1), i.e., *satisfying the inequalities* $\mu(t) \leq C_1\,\mu(s)\,\delta(s,t)^{N_1}$, *one has the estimate*

$$\|\varphi_\gamma\|_{k;\mu} \leq C(B,k,C_1,N_1)\,\mu(Pe)\,(1 + |b|_{Pe})^k.$$

PROOF In the case when $k = 0$, we must estimate the norm in $L^2(\Lambda)$ of $\mu\varphi_\gamma$ or, what amounts to the same, that of the function

$$t \mapsto \mu(Pt)\,\varphi(t). \tag{12.12}$$

When t lies in the support of φ, the distance $\delta(Pt,Pe) = \delta(t,e)$ remains bounded and since $\mu(Pt) \leq C_1\,\mu(Pe)\,\delta(e,t)^{N_1}$, the estimate is valid in that case. If k is a positive integer, we estimate $\|\varphi_\gamma\|_{k;\mu} = \|\mu\varphi_\gamma\|_k$ by means of Proposition 12.5. In the various terms involved to compute

$$(e_{P_1} \cdots e_{P_k} (\mu\varphi_\gamma))(t),$$

the only troublesome factors come from differentiating the exponential $e^{2i\pi\langle b,t\rangle}$. Now, if $e_p = \Sigma\, \alpha_j\, \partial/\partial t_j$ satisfies $\|e_p\|_t = 1$, one has $|\Sigma\, b_j\, \alpha_j| \leq |b|_t$ and one may conclude as before since $|b|_{Pt} \leq C(B)|b|_{Pe}$ whenever t lies in the support of φ, by Lemma 4.29. If $b \neq 0$, choose $\alpha \in \mathbb{R}^n$ with $\|\alpha\|_{Pe} = 1$ and $|b|_{Pe} = \langle b,\alpha\rangle$ and, setting $\alpha\, \partial/\partial t = \Sigma\alpha_j\, \partial/\partial t_j$ write

$$\begin{aligned} 2i\pi\, |b|_{Pe}\, \varphi_\gamma(t) &= \left(\alpha \frac{\partial}{\partial t}\right)\varphi_\gamma - \left[\left(\alpha \frac{\partial}{\partial t}\right)\varphi(P^{-1}t)\right] e^{2i\pi\langle b,t\rangle} \\ &= \left(\alpha \frac{\partial}{\partial t}\right)\varphi_\gamma - \psi_\gamma(t) \end{aligned} \tag{12.13}$$

with

$$\psi(s) = \left((P^{-1}\alpha)\frac{\partial}{\partial s}\right)\varphi(s). \tag{12.14}$$

As $|P^{-1}\alpha| = \|\alpha\|_{Pe} = 1$, the second term on the right-hand side of (12.13) remains bounded in $L^2(\Lambda)$ as φ varies in B and γ varies in G: since, as t lies in the support of φ, $\delta(t,Pe)$ is bounded, so is $\|\alpha\|_t$, which implies that the first term on the right-hand side remains bounded in H^{-1}. This yields the sought-after estimate in the case when $k = -1$ and $\mu = 1$, but arbitrary multiplicators μ can be taken care of in the same way as before, and iterating the procedure gives the result whenever k is a negative integer. We get it for any real number k as a consequence of the second assertion in Proposition 12.4. Q.E.D.

PROPOSITION 12.15
Let $\beta \in \mathbb{R}^n \setminus \{0\}$. *The function* μ_β *such that* $\mu_\beta(y) = |\beta|_y = |P_y^{1/2}\beta|$ *is a multiplicator. Let* $\|\ \|_{HS}$ *denote the Hilbert-Schmidt norm on* $M_n(\mathbb{R})$, *the space of* $n \times n$*-matrices with real coefficients. Given any* $A \in M_n(\mathbb{R})$, $A \neq 0$, *the function* μ_A *on* Λ *such that* $\mu_A(y) = \|P_y^{-1/2}\, A\, P_y^{1/2}\|_{HS}$ *is a multiplicator.*

PROOF That μ_β and μ_A are weight functions is a consequence of Lemma 4.29. Observe, moreover, that C_1 and N_1, defined as in (6.7), depend only on Λ. To show that μ_β is a symbol of weight μ_β, we may use (6.22). With $y = Pe$, $P \in GL(\Lambda)$, we have to show that

$$\left|\left(\frac{\partial}{\partial x}\right)^\alpha (|\beta|_{Px})(x = e)\right| \leq C(\alpha)\,|\beta|_y. \tag{12.16}$$

Now one has $|\beta|_{Px} = |P_x^{1/2}P'\beta|$ and $|P'\beta| = |\beta|_y$ so that the first estimate follows from the fact that $P_x^{1/2}$ depends smoothly on x. Given any $P \in GL(\Lambda)$ such that $Pe = y$, one has $\|P_y^{-1/2} A\, P_y^{1/2}\|_{HS} = \|P^{-1}\, AP\|_{HS}$, from which it follows that $\|P_{Px}^{-1/2}\, AP_{Px}^{1/2}\|_{HS} = \|P_x^{-1/2} P^{-1}\, APP_x^{1/2}\|_{HS}$. As a consequence, μ_A is a multiplicator. Q.E.D.

Let $\mathfrak{g} = \mathfrak{gl}(\Lambda) \times \mathbb{R}^n$ (semi-direct product) denote the Lie algebra of G. It consists of all pairs $T = (A,\beta)$ with $A \in \mathfrak{gl}(\Lambda)$ and $\beta \in \mathbb{R}^n$, and the exponential map is given by

$$e^{\epsilon T} = \left(e^{\epsilon A}, \frac{1 - e^{-\epsilon A'}}{A'}\beta\right)$$

[by (3.22), this is indeed a one-parameter subgroup of G]. Let

$$\frac{1}{2i\pi}\, dU(T) := \frac{1}{2i\pi}\frac{d}{d\epsilon}\, U(e^{\epsilon T})(\epsilon = 0) \tag{12.17}$$

denote the infinitesimal generator of the representation U defined in (3.24). We also abbreviate $dU(A,0)$ as $dU(A)$ and $dU(0,\beta)$ as $dU(\beta)$.

PROPOSITION 12.18
For $(A,\beta) \in \mathfrak{g}$ *and* $(P,b) \in G$, *we have*

$$\frac{1}{2i\pi}\, dU(A,\beta) = -\frac{1}{2i\pi}\,(At)\,\frac{\partial}{\partial t} + \langle\beta,t\rangle,$$

and the adjoint representation of G *on* $\mathfrak{g}$ *satisfies*

$$Ad(P,b)(A,\beta) = (PAP^{-1}, (PAP^{-1})'b + P'^{-1}\beta).$$

PROOF The first part is an immediate consequence of (3.24) and (12.17). For the second part, we let

$$(Ad(P, b))T = \left[\frac{d}{d\epsilon}\,(P, b)e^{\epsilon T}(P, b)^{-1}\right](\epsilon = 0) \tag{12.19}$$

and apply (3.22) as well as $(P,b)^{-1} = (P^{-1}, -P'b)$, which yields the result after a routine calculation. Q.E.D.

THEOREM 12.20
Given two linear operators A *and* B, *one from* $\mathcal{S}(\Lambda)$ *to* $\mathcal{S}'(\Lambda)$ *and the other both from* $\mathcal{S}(\Lambda)$ *to* $\mathcal{S}(\Lambda)$ *and from* $\mathcal{S}'(\Lambda)$ *to* $\mathcal{S}'(\Lambda)$, *we shall denote as* (ad A)(B) *the*

operator $[A,B] = AB - BA$. *Let* μ *be an arbitrary multiplicator and let* k *be a real number. The following two conditions on a linear operator* $A : \mathcal{S}(\Lambda) \to \mathcal{S}'(\Lambda)$ *are equivalent:*

(i) one has $A = \mathrm{Op}(f)$ *for some* $f \in S^k(\mu)$,

(ii) for any multiplicator ν, *any real number* j, *any set* $\{\beta_1, \ldots, \beta_p\}$ *in* $\mathbb{R}^n \backslash \{0\}$ *and any set* $\{A_1, \ldots, A_q\}$ *in* $\mathfrak{gl}(\Lambda)$, *the operator*

$$\mathrm{ad}(dU(\beta_1)) \ldots \mathrm{ad}(dU(\beta_p))\mathrm{ad}(dU(A_1)) \ldots \mathrm{ad}(dU(A_q))(A)$$

has an extension as a continuous operator from $H^j(\nu)$ *to* $H^{j-k+p}(\nu\mu^{-1}\mu_{\beta_1}^{-1} \ldots \mu_{\beta_q}^{-1} \mu_{A_1}^{-1} \ldots \mu_{A_q}^{-1})$. *Its bound as such does not depend on the choice of* $\{\beta_1, \ldots, \beta_p, A_1, \ldots, A_q\}$.

PROOF We first show that (i) implies (ii). According to Proposition 11.23, the symbol of $[dU(\beta), \mathrm{Op}(f)]$ is $-\langle \beta, \partial f/\partial\eta\rangle$ and that of $[dU(A), \mathrm{Op}(f)]$ is $(-(Ay)\,\partial/\partial y + (A'\eta)\,\partial/\partial\eta)f$. By Proposition 12.10, all we have to show is that if $g \in S^k(\mu)$ then $\langle \beta, \partial g/\partial\eta\rangle$ belongs to $S^{k-1}(\mu\mu_\beta)$ for all $\beta \in \mathbb{R}^n$ and $((Ay)\,\partial/\partial y - (A'\eta)\,\partial/\partial\eta)g$ belongs to $S^k(\mu\mu_A)$ for all $A \in \mathfrak{gl}(\Lambda)$; also, we must show that the estimates that characterize this fact are uniform with respect to the choice of β or A. At $y = Pe$, one may apply $P'\,\partial/\partial y$ once to a symbol of the type considered with no loss, or one may apply $P^{-1}\,\partial/\partial\eta$ and gain $(1 + |\eta|_y^2)^{-1/2}$. One gets the required estimates if one writes

$$\left\langle \beta, \frac{\partial g}{\partial\eta}\right\rangle = \left\langle P'\beta, P^{-1}\frac{\partial g}{\partial\eta}\right\rangle \tag{12.21}$$

and

$$\left((Ay)\frac{\partial}{\partial y} - (A'\eta)\frac{\partial}{\partial\eta}\right)g = \left\langle P^{-1}APe, P'\frac{\partial g}{\partial y}\right\rangle - \left\langle (P^{-1}AP)'P'\eta, P^{-1}\frac{\partial g}{\partial\eta}\right\rangle, \tag{12.22}$$

noting that $|P'\beta| = \mu_\beta(y)$ and that $|P'\eta| = |\eta|_y$. To estimate the derivatives, nothing new is involved. One should not forget, however, to "freeze" the coefficients of P' when iterating the operator $P'\,\partial/\partial y$; alternatively, one can use (6.22) again.

The proof that (ii) implies (i) is much more tricky. Let us observe, to start with, that it suffices to show that (ii) implies that $A = \mathrm{Op}(f)$ where f is a symbol of uniform type and weight m, with

$$m(y,\eta) = (1 + |\eta|_y^2)^{k/2}\,\mu(y). \tag{12.23}$$

Indeed, according to Definition 12.1, what has to be shown on top of that is only that any normalized $\partial/\partial\eta$-differentiation permits to gain the factor $(1 + |\eta|_y^2)^{-1/2}$, which is a consequence of (ii) together with the fact that $-\langle\beta, \partial f/\partial\eta\rangle$ is the symbol of $[dU(\beta), Op(f)]$. Thus, let us assume, in all that follows, that the operator A, as well as its formal adjoint as an operator from $\mathcal{S}(\Lambda)$ to $\mathcal{S}'(\Lambda)$, satisfies the following estimates [special cases of (ii)]:

$$\begin{aligned}\|((ad(dU(\beta_1)) \cdots ad(dU(\beta_p))ad(dU(A_1) \cdots ad(dU(A_q))A)u\|_{-k;\mu^{-1}} \\ \leq C\|u\|_{-p;\mu_{\beta_1}\cdots\mu_{\beta_p}\mu_{A_1}\cdots\mu_{A_q}}\end{aligned} \tag{12.24}$$

with constants C depending on A, k, μ, p, q but not on the set $\{\beta_1, \ldots, \beta_p; A_1, \ldots, A_q\}$. According to the characterization of operators of uniform type and given weight given in Theorems 9.19 and 10.1, what we have to show is that there exist constants C_N (depending also on A, k, μ, φ) such that the estimates

$$\begin{aligned}|(AU(P, b)\varphi\,|\,U(P_1, b_1)\varphi)| \leq C_N\,[(1 + |b|_{Pe})^k\mu(Pe) + (1 + |b_1|_{P_1e})^k\mu(P_1e)] \\ \delta(Pe, P_1e)^{-2N}(1 + |b - b_1|_{Pe})^{-N}.\end{aligned} \tag{12.25}$$

are valid. Here we may assume that φ is any fixed nonzero function in $C^\infty(\Lambda)$ with compact support. Also, we shall write

$$\gamma = (P,b), \quad \gamma_1 = (P_1,b_1) \tag{12.26}$$

and we shall abbreviate $U(\gamma)\psi$ as ψ_γ for any ψ. Let us observe, incidentally, that since, in (ii), ν and j are arbitrary, it is true that A acts as a continuous operator from $\mathcal{S}(\Lambda)$ to $\mathcal{S}(\Lambda)$. Starting from

$$|(A\varphi_\gamma\,|\,\varphi_{\gamma_1})| \leq C\|A\varphi_\gamma\|_{-k;\mu^{-1}}\|\varphi_{\gamma_1}\|_{k;\mu} \leq C\|\varphi_\gamma\|_{L^2(\Lambda)}\|\varphi_{\gamma_1}\|_{k;\mu}\,, \tag{12.27}$$

we get the case N = 0 of (12.25) as an immediate consequence of Lemma 12.11. We still have to show how to improve the estimate [using the full force of (12.24)] so as to gain arbitrary powers of $\delta(Pe, P_1e)^{-1}$ or arbitrary powers of $(1 + |b - b_1|_{Pe})^{-1}$, in the latter case at the possible loss of some power of $\delta(Pe, P_1e)$. This requires a lemma, in which we use the notation introduced in Definition 12.17.

LEMMA 12.28

For each subset σ *of* $\{1, \ldots, N\}$, *denote points in* σ *as* $i_1, \ldots, i_s$ *with* $N \geq i_1 > \ldots > i_s \geq 1$ *and points in the complementary of* σ *as* $j_1, \ldots, j_r$ *with*

$r + s = N$ *and* $1 \leq j_1 < \ldots < j_r \leq N$; *let* $\epsilon(\sigma) = (-1)^r$. *Then, for any pair* (φ,ψ) *of functions in* $C_0^\infty(\Lambda)$, *any sequence* $\{T_1, \ldots, T_N\}$ *in* $\mathfrak{g}$ *and any* $(\gamma,\gamma_1) \in G \times G$, *one has the identity*

$$(A\varphi_\gamma \,|\, dU(Ad(\gamma)T_N) \cdots dU(Ad(\gamma)T_1)\psi_{\gamma_1}) = \sum_\sigma \epsilon(\sigma)$$

$$\cdot ([\ldots[A, dU(Ad(\gamma)T_{i_1})], \ldots, dU(Ad(\gamma)T_{i_s})](dU(T_{j_1})\cdots dU(T_{j_r})\varphi)_\gamma \,|\, \psi_{\gamma_1}).$$

PROOF In the case when $N = 1$ and $T_1 = T$ for simplicity of notation, it suffices to write

$$dU(Ad(\gamma)T)A\varphi_\gamma + [A, dU(Ad(\gamma)T)]\varphi_\gamma$$
$$= A\, dU(Ad(\gamma)T)\varphi_\gamma = A\, U(\gamma)dU(T)U(\gamma^{-1})\varphi_\gamma = A(dU(T)\varphi)_\gamma$$

and to remember that $(2\pi i)^{-1}dU(Ad(\gamma)T)$ is a self-adjoint operator. The general case follows by induction. Q.E.D.

End of proof of Theorem 12.20. According to (4.11), to gain powers of $\delta(Pe,P_1e)^{-2}$ amounts to gaining powers of $\|P^{-1}P_1\|^{-1}$ since the assumption (ii) is invariant under the substitution $A \mapsto A^*$. Given P and $P_1 \in GL(\Lambda)$, choose $\beta_1 \in \partial\Lambda$, the boundary of Λ, so that $|\beta_1| = 1$ and $|(P_1^{-1}P)'\beta_1| = \|P^{-1}P_1\|^{-1}$. This is possible in view of Lemma 4.64, and we set $\beta = (P_1^{-1}P)'\beta_1$. Also, we set $\psi(t) = \langle\beta_1, t\rangle^{-N}\,\varphi(t)$, so that ψ describes (as β_1 varies) a bounded subset of $C_0^\infty(\Lambda)$. Finally, we apply Lemma 12.28 with $T_j = T = (0,\beta)$ for all j. Proposition 12.18 shows that $(2\pi i)^{-1}dU(Ad(\gamma)T)$ is the operator of multiplication by $\langle P'^{-1}\beta, t\rangle$: since $P'^{-1}\beta = P_1'^{-1}\beta_1$, the left-hand side

$$(2i\pi)^N\, (A\varphi_\gamma \,|\, \langle P'^{-1}\beta,t\rangle^N\, \langle\beta_1, P_1^{-1}t\rangle^{-N}\, \varphi(P_1^{-1}t)\, e^{2i\pi\langle b_1,t\rangle})$$

of the identity asserted by Lemma 12.28 reduces to $(2i\pi)^N\,(A\varphi_\gamma|\varphi_{\gamma_1})$. For any given term on the right-hand side, let $\chi = (dU(0,\beta))^r\varphi$. We wish to show that (with $r + s = N$)

$$|(((ad\, dU(Ad(\gamma)T))^s A)\chi_\gamma \,|\, \psi_{\gamma_1})|$$
$$\leq C\|P^{-1}P_1\|^{-N}\mu\,(P_1e)(1 + |\, b_1\,|_{P_1e})^k \quad (12.29)$$

and Lemma 12.11, together with Schwarz' inequality, makes it enough to show that

$$\|((\text{ad } dU(\text{Ad}(\gamma)T))^s A)\chi_\gamma\|_{-k;\mu^{-1}} \leq C\|P^{-1}P_1\|^{-N}. \tag{12.30}$$

By (12.24), it suffices to show that

$$\|\chi_\gamma\|_{-1;\mu^s_{P'^{-1}\beta}} \leq C\,\|P^{-1}P_1\|^{-N}, \tag{12.31}$$

an inequality in which the subscript -1 may of course be replaced by 0 [the full force of (12.24) shall be needed later]. Now

$$\mu^s_{P'^{-1}\beta}\,(Pe) = |P'^{-1}\beta|^s_{Pe} = |\beta|^s = \|P^{-1}P_1\|^{-s}, \tag{12.32}$$

and, since χ is $\|P^{-1}P_1\|^{-r}$ times a function that remains in a bounded subset of $C_0^\infty(\Lambda)$, (12.31) is a consequence of (12.32) and Lemma 12.11. We have thus far proved that

$$|(A\varphi_\gamma|\varphi_{\gamma_1})| \leq C_N\,[(1 + |b|_{Pe})^k\mu(Pe) + (1 + |b_1|_{P_1e})^k\,\mu(P_1e)]\,\delta(Pe,P_1e)^{-N}. \tag{12.33}$$

It remains to be shown how to gain arbitrary powers of $(1 + |b - b_1|_{Pe})^{-1}$ at a possible loss of powers of $\delta(Pe,P_1e)$. We consider only the first power $(1 + |b - b_1|_{Pe})^{-1}$, leaving it to the reader to apply the general case of Lemma 12.28. This time, we take $T = (A,0)$ so that Proposition 12.18 implies

$$dU(\text{Ad}(\gamma)A) = -\,(PAP^{-1}t)\,\frac{\partial}{\partial t} + 2\pi i\,\langle (PAP^{-1})'b,\,t\rangle. \tag{12.34}$$

As

$$\psi_{\gamma_1}(t) = \psi(P_1^{-1}t)\,e^{2i\pi\langle b_1,t\rangle}, \tag{12.35}$$

one has

$$(dU(\text{Ad}(\gamma)A)\psi_{\gamma_1})(t) = -\left((P_1^{-1}PAP^{-1}P_1s)\,\frac{\partial}{\partial s}\,\psi\right)(s = P_1^{-1}t)\,e^{2i\pi\langle b_1,t\rangle} + 2\pi i\,\langle (PAP^{-1})'(b - b_1),\,t\rangle\,\psi_{\gamma_1}(t) \tag{12.36}$$

or

$$(dU(\text{Ad}(\gamma)A)\psi_{\gamma_1})(t) = -\left((P_1^{-1}PAP^{-1}P_1t)\,\frac{\partial}{\partial t}\,\psi\right)_{\gamma_1} + 2\pi i\,(\langle P_1'P'^{-1}A'P'(b - b_1),\,t\rangle\psi)_{\gamma_1}. \tag{12.37}$$

We may assume that $|b - b_1|_{Pe} \geq 1$. According to Lemma 4.65, it is possible to find $A_1 \in \mathfrak{gl}(\Lambda)$ such that, with $\theta = A_1{}'P'(b - b_1)$, the following three conditions are valid: $|\theta| = 1$, θ lies on the boundary of Λ, and $\|A_1\| \leq C|P'(b - b_1)|^{-1} = C|b - b_1|_{Pe}^{-1}$. Now $\theta_1 = P'_1P'^{-1}\theta$ satisfies $|\theta_1| \geq \|P'P'^{-1}_1\|^{-1} = \|P_1^{-1}P\|^{-1}$ and, for a fixed $\varphi \in C_0^\infty(\Lambda)$, the function $\|P_1^{-1}P\|^{-1}$ $(\langle\theta_1, t\rangle)^{-1}\,\varphi$ varies in a bounded subset of $C_0^\infty(\Lambda)$ as γ and γ_1 describe the group G (since $|\theta_1| \leq C(\varphi)\,\langle\theta_1,t\rangle$ on the support of φ). We now apply the fundamental identity expressed by Lemma 12.28 (or rather its special case $N = 1$, with $T_1 = T$) with $\psi(t) = \langle\theta_1, t\rangle^{-1}\varphi(t)$. By (12.37), one has

$$dU(Ad(\gamma)A_1)\psi_{\gamma_1} = -\left((P_1^{-1}PA_1P^{-1}P_1t)\,\frac{\partial}{\partial t}\,\psi\right)_{\gamma_1} + 2\pi i\,\varphi_{\gamma_1} \qquad (12.38)$$

so that Lemma 12.28 may be expressed as

$$2\pi i(A\varphi_\gamma\,|\,\varphi_{\gamma_1}) = \left(A\varphi_\gamma\,\Big|\left((P_1^{-1}PA_1P^{-1}P_1t)\,\frac{\partial}{\partial t}\,\psi\right)_{\gamma_1}\right)$$
$$- ([A, dU(Ad(\gamma)A_1]\varphi_\gamma\,|\,\psi_{\gamma_1}) + (A(dU(A_1)\varphi)_\gamma\,|\,\psi_{\gamma_1}). \qquad (12.39)$$

Recall that $\|A_1\| \leq C\,|b - b_1|_{Pe}^{-1}$. From what was said above, the function $\delta(Pe,P_1e)^{-2}\psi$ remains in a bounded subset of $C_0^\infty(\Lambda)$. Applying Schwarz' inequality together with the case $p = q = 0$ of (12.24) on the left, and Lemma 12.11 on the right, it is easy to show that the first and third terms on the right-hand side of (12.39) are bounded by $C\,|b - b_1|_{Pe}^{-1}\,\delta(Pe,P_1e)^6$. Finally, the second term on the right-hand side of (12.39) may be written as

$$([A, dU(Ad(\gamma)A_1)]\varphi_\gamma\,|\,\psi_{\gamma_1}) = ([A, dU(PA_1P^{-1})]\varphi_\gamma\,|\,\psi_{\gamma_1})$$
$$+ ([A, dU((PA_1P^{-1})'b)\varphi_\gamma\,|\,\psi_{\gamma_1}) \qquad (12.40)$$

so that (12.24) yields

$$|([A, dU(Ad(\gamma)A_1)]\varphi_\gamma\,|\,\psi_{\gamma_1})|$$
$$\leq C(\|\varphi_\gamma\|_{0;\mu_{PA_1P^{-1}}} + \|\varphi_\gamma\|_{-1;\mu_{(PA_1P^{-1})'b}})\|\psi_{\gamma_1}\|_{k;\mu} \qquad (12.41)$$

where we bound again $\|\psi_{\gamma_1}\|_k$ by Lemma 12.11. Applying Lemma 12.11 to the two terms within parentheses too, we are led to write

$$\mu_{PA_1P^{-1}}(Pe) = \|P_{Pe}^{-1/2}PA_1P^{-1}P_{Pe}^{1/2}\|_{HS} = \|A_1\|_{HS} \qquad (12.42)$$

since the matrix $P_{Pe}^{-1/2}P$ is orthogonal, and

$$\mu_{(PA_1P^{-1})'b}(Pe)(1 + |b|_{Pe})^{-1} = |P'^{-1}A_1'P'b|_{Pe}(1 + |P'b|)^{-1}$$
$$= |A_1'P'b|(1 + |P'b|)^{-1} \leq \|A_1\|_{HS}. \quad (12.43)$$

This concludes the proof of Theorem 12.20. Q.E.D.

13

Action of Diffeomorphisms on Operators of Classical Type

The diffeomorphisms $\Phi : \Lambda \to \Lambda$ we are concerned with in this section are global. It would be an easy matter, however, to localize everything (both Φ and the operators) near the vertex of Λ. In this way, one should be able to define and study operators of classical type on any manifold X modeled on a symmetric cone (actually a half-space everywhere save at exceptional points) near each point of the topological boundary ∂X. However, the piecing together, in a suitable C^∞ sense, of the pseudodifferential analyses related to parts of X modeled on two different cones cannot be a trivial task in view of the stratified structure (in general) of the boundary of a convex symmetric cone. The diffeomorphism Φ must preserve some of the structure of Λ: in particular, the norms $\| \ \|_t$ and $\| \ \|_{\Phi(t)}$ should be comparable in a uniform way, which is best expressed with the help of Lemma 4.29; besides, bounds on the derivatives are required.

DEFINITION 13.1 Let $\Phi : \Lambda \to \Lambda$ be a C^∞ diffeomorphism with inverse Ψ. We shall say that Φ is admissible if the following three conditions are satisfied:
(i) for some constant C_1, one has $\delta(t, \Phi(t)) \leq C_1$ for all $t \in \Lambda$;
(ii) given $\beta \in \mathbb{R}^n$, the function $(y,\eta) \mapsto \langle \Phi(y), \beta \rangle$ (depending only on y) is a symbol of uniform type and weight $|\beta|_y$, in a way that is uniform, too, with respect to β;
(iii) same as (ii) with Φ replaced by Ψ.

Remarks
1) what makes (ii) slightly more cumbersome than deserved is that we have not discussed vector-valued symbols (a trivial but paper-consuming task); this will also complicate the next lemma in a mild way; 2) in the case when $\Lambda = \mathbb{R}_+$, Φ is admissible if and only if estimates of the kind

$$|(t\frac{d}{dt})^j\, \Phi(t)| \leq C(j)\ t$$

are valid, and if the same is true with Φ replaced by $\Psi = \Phi^{-1}$.

LEMMA 13.2
Let $\Phi : \Lambda \to \Lambda$ be an admissible diffeomorphism. As P describes $GL(\Lambda)$, the components of the vector $P^{-1}\Phi(Px)$ have bounded derivatives of all orders at $x = e$, in a way that depends on the order of differentiation but is uniform with respect to P. In the same way, the entries of the matrix $P^{-1}d\Phi(Px)P$, as well as the components of the (1-contravariant and 2-covariant) tensor $P^{-1}d^2\Phi(Px)(P \otimes P)$, have bounded derivatives of all orders at $x = e$.

PROOF By (6.22), to say that $\langle\Phi(y), \beta\rangle$ is of uniform type and weight $|\beta|_y$ amounts to saying that the derivatives of all orders, at $x = e$, of the function $x \mapsto \langle\Phi(Px),\beta\rangle$, can be majorized by $|\beta|_{Pe} = |P'\beta|$ in a uniform way. In other words, the components of the vector $P^{-1}\Phi(Px)$ have bounded derivatives of all orders at $x = e$ (in a way uniform with respect to P): this proves the first assertion. Since the mapping $x \mapsto P^{-1}\Phi(Px)$ has the first and second derivatives

$$d(P^{-1}\Phi(Px))h = P^{-1}d\Phi(Px)Ph \qquad (h \in \mathbb{R}^n)$$

and

$$d^2(P^{-1}\Phi(Px))(h \otimes k) = P^{-1}d^2\Phi(Px)(Ph \otimes Pk) \qquad (h,k \in \mathbb{R}^n),$$

the remaining assertions follow. Q.E.D.

PROPOSITION 13.3
Let $\Phi : \Lambda \to \Lambda$ be an admissible diffeomorphism. Then the map $u \mapsto v = u \circ \Phi$ preserves (as topological vector spaces) all Sobolev spaces H^k, as well as the space $\mathcal{S}(\Lambda)$.

PROOF Since $\omega(t) = \det P_t^{1/2}$, condition (i), Lemma 13.2 and Lemma 6.10 show that if $\Phi_*(dm)$ denotes the image of dm under Φ, the density of $\Phi_*(dm)$ with respect to dm lies between two fixed positive bounds: thus the topological vector space $L^2(\Lambda)$ is preserved under the map under consideration. Using duality and an interpolation argument (Proposition 12.4), we may, without loss of generality, show that H^k is preserved only in the case when k is a positive integer. In this case we may use Proposition

12.5 and the notations there; denoting the constant vector field e_p as $\alpha\, \partial/\partial t$ and letting $\alpha = P_t^{1/2}\beta$ (so that $\|\alpha\|_t = |\beta|$), let

$$\Phi_*\left(\alpha \frac{\partial}{\partial t}\right) = (d\Phi(\Phi^{-1}(t))\alpha)\frac{\partial}{\partial t} = ((d\Psi)^{-1}(t)\alpha)\frac{\partial}{\partial t} \tag{13.4}$$

be the image of $\alpha\, \partial/\partial t$ under Φ. Since

$$du(Pe)P\beta = d(u \circ P)(e)\beta \tag{13.5}$$

we have

$$\Phi_*\left(\alpha \frac{\partial}{\partial t}\right)u(t) = \left(((P^{-1}\, d\Psi(Px)P)^{-1}\beta)\frac{\partial}{\partial x}\right)(u \circ P)(x = e). \tag{13.6}$$

The fact that $u \circ \Phi \in H^1$ if $u \in H^1$ is then a consequence of Proposition 12.5 together with the estimates for the entries of the matrix $P^{-1}d\Psi(Px)P$ provided by Lemma 13.2. The same lemma makes it possible to apply to u the composition of several operators of the type defined in (13.6), which concludes the proof of the first part of Proposition 13.3. To show that the space $\mathcal{S}(\Lambda)$ is preserved as well under the map $u \mapsto u \circ \Phi$, we just have to trace the effect of that map on weights that are polynomials in t or powers of $\Delta(t)$: in the second case, this is taken care of by Lemma 6.10; finally, that $|\Psi(t)| \leq C_1|t|$ for some constant C_1 is a consequence of the inequality $\|P_t\| \leq C|t|^2$ (that follows from (4.30)) together with condition (iii) in Definition 13.1. This concludes the proof of Proposition 13.3.

Q.E.D.

PROPOSITION 13.7

Let Φ be an admissible diffeomorphism of Λ. If μ is a multiplicator in the sense of Definition 12.1, so is $\mu \circ \Phi$: if, then, $u \in H^k(\mu)$, it follows that $u \circ \Phi \in H^k(\mu \circ \Phi)$. If μ has the form μ_β or μ_M introduced in Proposition 12.15, then (in a way that is uniform with respect to β or M) the spaces $H^k(\mu)$ and $H^k(\mu \circ \Phi)$ coincide, for all k, as topological vector spaces.

PROOF Condition (i) in Definition 13.1 and Lemma 4.12 show that $\mu \circ \Phi$ is a weight-function if μ is a multiplicator: the same proof as the one of Proposition 13.3 (replacing, however, L^2-estimates by L^∞-estimates) then shows that the derivatives of $\mu \circ \Phi$ satisfy the estimates that make $\mu \circ \Phi$ a multiplicator. That $u \circ \Phi$ belongs to $H^k(\mu \circ \Phi)$ if $u \in H^k(\mu)$ is an immediate consequence of Definition 12.9 and of Proposition 13.3. Recalling that $\mu_\beta(y) = |\beta|_y$ and that $\mu_M(y) = \|P_y^{-1/2}MP_y^{1/2}\|_{HS}$, it follows from Definition

13.1 and from Lemma 4.29 that if μ has one of these two forms, then μ and $\mu \circ \Phi$ are equivalent as weight functions (in a way uniform with respect to b or M), in the sense that the function $\mu^{-1}(\mu \circ \Phi)$ lies between two fixed positive bounds. Finally, two multiplicators μ and ν equivalent as weight functions are also equivalent in the stronger sense that $(\mu\nu^{-1})^{\pm 1}$ is a multiplicator of weight one: then, for every $k \in \mathbb{R}$, the topological vector spaces $H^k(\mu)$ and $H^k(\nu)$ coincide, which concludes the proof of Proposition 13.7. Q.E.D.

DEFINITION 13.8 Given an admissible diffeomorphism $\Phi : \Lambda \to \Lambda$ and a linear operator $A : \mathcal{S}(\Lambda) \to \mathcal{S}(\Lambda)$, we denote as $\Phi_* A$ the operator characterized by

$$(\Phi_* A)u = (A(u \circ \Phi)) \circ \Phi^{-1}$$

for every $u \in \mathcal{S}(\Lambda)$.

Our aim in this section is to show that if $A = \mathrm{Op}(f)$ for $f \in S^k(\mu)$, then $\Phi_* A = \mathrm{Op}(g)$ for some $g \in S^k(\mu \circ \Phi^{-1})$. To this effect, we shall show that $\Phi_* A$ satisfies condition (ii) of Theorem 12.20. With dU(T) as in Definition 12.17, and $\Psi = \Phi^{-1}$, we note that

$$[dU(T), \Phi_* A] = \Phi_*([\Psi_* dU(T), A]), \tag{13.9}$$

which is the reason why we have to study the operators $\Psi_* dU(\beta)$ and $\Psi_* dU(M)$, for any $\beta \in \mathbb{R}^n \backslash \{0\}$ or $M \in \mathfrak{gl}(\Lambda)$.

LEMMA 13.10

Let $\Phi : \Lambda \to \Lambda$ be an admissible diffeomorphism: with inverse Ψ. For every $\beta \in \mathbb{R}^n \backslash \{0\}$, the symbol of the operator $\Psi_* dU(\beta)$ lies in the class $S^0(\mu_\beta)$ in the sense of Definition 12.1; for every $M \in \mathfrak{gl}(\Lambda)$ the (active) symbol of the operator $\Psi_* dU(M)$ lies in $S^1(\mu_M)$. All estimates are uniform with respect to β or M.

PROOF Since by Proposition 12.18, $dU(M) = -(Mt)\, \partial/\partial t$ and $dU(\beta)$ is the operator of multiplication by $2\pi i\langle \beta, t\rangle$, it is immediate that $\Psi_* dU(\beta)$ is the operator of multiplication by $2\pi i\langle \beta, \Phi(t)\rangle$: the symbol of that operator is the function $(y,\eta) \mapsto 2\pi i\langle \beta, \Phi(y)\rangle$, and Definition 13.1 implies that this is (in a way uniform with respect to β), a symbol of uniform type and weight $|\beta|_y = \mu_\beta(y)$; as this symbol does not depend on η, we may as well say that it lies in the class $S^0(\mu_\beta)$ (cf. Definition 12.1). On the other hand,

$$(\Psi_* dU(M))u = -\left(t \mapsto (Mt)\,\frac{\partial}{\partial t}\,(u \circ \Psi)\right) \circ \Phi$$

$$= -\left[t \mapsto \sum M_{jk} t_k u'_\ell(\Psi(t))\,\frac{\partial \Psi_\ell}{\partial t_j}\right] \circ \Phi \tag{13.11}$$

so that

$$\Psi_* dU(M) = -\sum a_\ell \frac{\partial}{\partial t_\ell} \tag{13.12}$$

with

$$a_\ell = \sum_{jk} \frac{\partial \Psi_\ell}{\partial s_j}\,(s = \Phi(t)) M_{jk}(\Phi(t))_k\,. \tag{13.13}$$

Let A be the operator with active symbol $\Sigma a_j(y)\eta_j$. By Proposition 11.18, together with the fact that the derivative at $t = s$ of the map $t \mapsto \mathrm{mid}(s,t)$ is half the identity, one has (say, for every $u \in \mathcal{S}(\Lambda)$)

$$2i\pi(Au)(s) = \sum a_j(s)\,\frac{\partial u}{\partial s_j}$$

$$+ \frac{1}{2}\left(\sum \frac{\partial a_j}{\partial s_j}\right) u(s) + \sum a_j(s)\varphi_j(s)u(s) \tag{13.14}$$

with

$$\varphi_j(s) = 2^n\,\frac{\partial}{\partial t_j}\,\mathrm{Det}\left(\frac{\partial}{\partial t}\,\mathrm{mid}(s,\,t)\right)(t = s). \tag{13.15}$$

Even though it is not necessary to compute $\Sigma a_j(s)\varphi_j(s) = \langle a,\varphi\rangle(s)$ to a full extent, one may use the same computation as in (11.22) to find (choosing $P = P_s^{1/2}$)

$$\langle a,\,\varphi\rangle(s) = -\frac{1}{2}\left((P_s^{-1/2}a)\,\frac{\partial}{\partial t}\right)(\det M_t)(t = e). \tag{13.16}$$

Thus the symbol of $\Psi_* dU(M)$ is the function

$$f(y,\,\eta) = -2\pi i\,\langle a(y),\,\eta\rangle$$

$$+ \frac{1}{2}\sum \frac{\partial a_j}{\partial y_j} - \frac{1}{2}\left((P_y^{-1/2}a)\,\frac{\partial}{\partial t}\right)(\det M_t)(t = e). \tag{13.17}$$

We now examine the three terms on the right-hand side of (13.17) separately, noting that the vector-valued function $a = (a_1, \ldots, a_n)$ may be written as

$$a(y) = d\Psi(\Phi(y))M\Phi(y) = (d\Phi(y))^{-1}M\Phi(y). \tag{13.18}$$

To show first that $\langle a(y), \eta\rangle$ is a symbol in $S^1(\mu_M)$ (recalling that $\mu_M(y) = \|P_y^{-1/2}MP_y^{1/2}\|_{HS}$) what we have to do, according to (6.22), is to examine, with $y = Pe$, the ordinary derivatives at $(x,\xi) = (e,0)$ of the function

$$\begin{aligned}\langle a(Px), P'^{-1}\xi + \eta\rangle &= \langle (d\Phi(Px))^{-1}M\Phi(Px), P'^{-1}\xi + \eta\rangle \\ &= \langle P^{-1}(d\Phi(Px))^{-1}P\cdot P^{-1}MP\cdot P^{-1}\Phi(Px), \xi + P'\eta\rangle. \end{aligned}\tag{13.19}$$

According to Lemma 13.2, the components of the vector $P^{-1}\Phi(Px)$, as well as the entries of the matrix $P^{-1}(d\Phi(Px))^{-1}P = (P^{-1}d\Phi(Px)P)^{-1}$, have bounded derivatives of all orders at $x = e$. Also, $\|P^{-1}MP\|_{HS} = \mu_M(y)$, and $|P'\eta| = |\eta|_y$. Thus $\langle a(y), \eta\rangle$ does indeed lie in $S^1(\mu_M)$ in a uniform way with respect to $M \in \mathfrak{gl}(\Lambda)$. We must show now that the last two terms on the right-hand side of (13.17) lie in $S^0(\mu_M)$. For the last one, we have to give bounds at $x = e$ for the derivatives of the function

$$\begin{aligned}&\left((P_{Px}^{-1/2}a(Px))\frac{\partial}{\partial t}\right)(\det M_t)(t = e)\\ &= \left((P_{Px}^{-1/2}(d\Phi(Px))^{-1}M\Phi(Px))\frac{\partial}{\partial t}\right)(\det M_t)(t = e)\\ &= \langle (P^{-1}d\Phi(Px)P)^{-1}\cdot P^{-1}MP\cdot P^{-1}\Phi(Px), \xi\rangle \end{aligned}\tag{13.20}$$

with

$$\xi = P'P_{Px}^{-1/2}\frac{\partial}{\partial t}(\mathrm{Det}\ M_t)(t = e). \tag{13.21}$$

Lemma 13.2 again gives the proper estimates for the derivatives of the vector on the left-side within brackets. Since, for some $k \in K$ (depending on P and x), one has $P_{Px}^{1/2} = PP_x^{1/2}k = k^{-1}P_x^{1/2}P'$ and the vector $\partial/\partial t\ (\mathrm{Det}\ M_t)(t = e)$ is K-invariant, we have

$$\xi = P_x^{-1/2}k\frac{\partial}{\partial t}(\mathrm{Det}\ M_t)(t = e) = P_x^{-1/2}\frac{\partial}{\partial t}(\mathrm{Det}\ M_t)(t = e), \tag{13.22}$$

so we are done, so far as the last term on the right-hand side of (13.17) is concerned. Finally, we compute the middle term, i.e.,

$$b(y) = \sum \frac{\partial a_\ell}{\partial y_\ell} = \sum \frac{\partial}{\partial y_\ell}\left(\frac{\partial \Psi_\ell}{\partial s_j}(s = \Phi(y))\right) M_{jk}\Phi_k(y)$$

$$+ \sum \frac{\partial \Psi_\ell}{\partial s_j}(\Phi(y)) M_{jk}\frac{\partial \Phi_k}{\partial y_\ell}, \tag{13.23}$$

where the second term on the right-hand side reduces to Tr(M); to show that the first term $b_1(y)$, too, lies in $S^0(\mu_M)$, we may as well, thanks to Proposition 13.7, consider instead $b_1 \circ \Psi$. Now

$$(b_1 \circ \Psi)(y) = \sum \frac{\partial \Phi_r}{\partial t_\ell}(t = \Psi(y)) \frac{\partial^2 \Psi_\ell}{\partial y_r \partial y_j} M_{jk} y_k$$

$$= d\Phi(\Psi(y)) \cdot (d^2\Psi)(y) \cdot My \tag{13.24}$$

where the first dot stands for the contraction of the indices ℓ, r and the second one for the contraction of j. We must show that the function

$$(d\Phi \circ \Psi)(Px) \cdot (d^2\Psi)(Px) \cdot MPx$$

has derivatives of all orders in x, at x = e, bounded by $C(\Psi)\, \|P^{-1}MP\|_{HS}$: this follows from Lemma 13.2 again since it can also be written as

$$P^{-1}(d\Phi \circ \Psi)(Px)P \cdot P^{-1}d^2\Psi(Px)(P \otimes P) \cdot P^{-1}MPx.$$

This concludes the proof of Lemma 13.10. Q.E.D.

THEOREM 13.25
Let $\Phi : \Lambda \to \Lambda$ be an admissible diffeomorphism: let $k \in \mathbb{R}$ and let μ be a multiplicator. For each symbol f in $S^k(\mu)$, the operator $\Phi_*\mathrm{Op}(f)$ may be written as Op(g) for some $g \in S^k(\mu \circ \Phi^{-1})$.

PROOF Again, let $\Psi = \Phi^{-1}$, and let A = Op(f). We want to show that Φ_*A satisfies condition (ii) of Theorem 12.16 (with $\mu \circ \Psi$ substituted for μ and, if one cares for it, $\nu \circ \Psi$ substituted for ν as well since ν is arbitrary). Now, from Proposition 13.7, it follows that if an operator B extends as a continuous linear operator from $H^j(\nu)$ to some space $H^i(\lambda)$, then Φ_*B extends as a continuous operator from $H^j(\nu \circ \Psi)$ to $H^i(\lambda \circ \Psi)$. Applying (13.9), one sees that all that has to be done is to show that A itself satisfies (ii)′, which is just the same as (ii) except that $dU(\beta_j)$ [resp. $dU(M_j)$] has to be replaced by $\Psi_* dU(\beta_j)$ [resp. $\Psi_* dU(M_j)$] for all j. By Proposition 12.2, the symbol of [X, A] belongs to $S^{k+i-1}(\mu\lambda)$ if the symbol of X (resp. A) belongs to $S^i(\lambda)$ [resp. $S^k(\mu)$]. Then (ii)′ is a consequence of Lemma 13.10 together with Proposition 12.10, which concludes the proof of Theorem 13.25. Q.E.D.

Before we leave, at this point, the pseudodifferential analysis on Λ to turn to questions more directly related to the complex analysis on Π, it may be useful to describe a few simple examples in a more concrete way. The remarks that follow are aimed at stressing the flexibility of the Fuchs calculus; they are not meant, however, even as an introduction to its applications. The subject of differential operators on some open subset of $\mathbb{R}^n$ which degenerate in a prescribed way at the boundary has a long-standing tradition (cf. [BG]). As a pseudodifferential analysis on domains with boundaries, the Fuchs calculus permits to handle simultaneously full algebras of operators modeled on differential operators with a specific degeneracy at the boundary: starting with $\mathbb{R}^+ = \{s : s > 0\}$, it is fair in this case, if meaningless in any strict sense, to regard operators of the Fuchs calculus as "functions" of the (non-commuting) operators s and $(1/2\pi i)\, s\, d/ds$, in the same way as operators of the Weyl calculus on the real line may be regarded as "functions" of the operators x and $(1/2\pi i)\, d/dx$; also, the specificity of the Fuchs type microlocal analysis at $s = 0$ lies in that one is not only interested in how differentiable a function $u = u(s)$ is near that point, but also in how flat it is. Recall from Definition 12.1 that when $\Lambda = \mathbb{R}^+$ the class of symbols $S^k(\mu)$, where μ is any multiplicator, can be characterized by the inequalities

$$\left|\left(y\frac{\partial}{\partial y}\right)^p\left(y^{-1}\frac{\partial}{\partial \eta}\right)^q f(y,\eta)\right| \leq C(p,q)\mu(y)(1 + y^2\eta^2)^{(k-q)/2}. \quad (13.26)$$

In particular, when $k \in \mathbb{N}$, $S^k = S^k(1)$ contains the class of (active or passive) symbols of differential operators A of order k of the kind

$$A = \sum_{j=0}^{k} b_j(s)\left(\frac{1}{2\pi i}\, s\, \frac{d}{ds}\right)^j \quad (13.27)$$

whose coefficients b_j satisfy the estimates $|(y\, d/dy)^p\, b_j(y)| \leq c(j,p)$ for $y > 0$: this is a consequence of the formula in Proposition 11.18; it is also reported with more details in ([U5; (7.6) and (7.7)], observing that the operator denoted Q(g) there would be denoted Op(f) here provided that the identity $f(y,\eta) = g(y\eta, y)$ holds. A more restricted class of differential operators A is obtained if one demands moreover that the coefficients b_j extend as C^∞ functions on $[0,\infty[$. The (passive or active) symbol of such an operator may be written as

$$\sum_{j=0}^{k} a_j(y)(y\eta)^j, \tag{13.28}$$

where the coefficients a_j satisfy exactly the same assumptions as the b_j's. Under the (ellipticity) assumption that $b_k(s)$ ($= a_k(s)$) does not vanish for $s \geq 0$, the usual method of inversion (similar to the proof of Proposition 12.3) permits to build parametrices (i.e., approximate right inverses) of A up to an error term in S^{-N} near $y = 0$, where N is as large as one pleases. However, as shown by the example $A = (1/2i\pi)\ s\ d/ds$, one cannot in general find any right inverse of A up to an error term in $S^0(y^\epsilon)$, $\epsilon > 0$, for this would entail the validity of the estimate $\|u\| \leq c\ \|Au\|$ whenever u has its support in $\{s : 0 < s \leq s_0\}$, s_0 small enough, and this is not possible. However, introduce the zeros $r_1, \ldots, r_k$ of the polynomial $\Sigma b_j(0)\ r^j$: it was then shown in ([U5; Section 9] that one can find right inverses of A up to error terms which lie in $S^{-N}(y^N)$, near $y = 0$, up to any given order of differentiability, provided that the distance of the set $\{ir_1, \ldots, ir_k\} \subset \mathbb{C}^k$ to $\mathbb{R}^k$ is large enough. The condition $b_k(0) \neq 0$ just means, in the language of ordinary differential equations theory, that the equation $Au = v$ ($v \in C^\infty([0,\infty[)$) is of Fuchs type at $s = 0$. The preceding simple example was at the origin of the terminology adopted for the Fuchs calculus.

After the real line, the next simplest case is the half-space $\mathbb{R}^{n+1}_+ = \mathbb{R}_+ \times \mathbb{R}^n = \{(s,x) : s > 0\}$, a local model for manifolds with smooth boundary. In this case, it is natural to use the "tensor" product of the Fuchs calculus in the s variable and the Weyl calculus in the x-variables: the basic Hilbert space is therefore $L^2(\mathbb{R}^{n+1}_+; s^{-1}ds\ dx)$, symbols $f = f(s,x; \sigma,\xi)$ live on $\mathbb{R}_+ \times \mathbb{R}^n \times \mathbb{R} \times \mathbb{R}^n$, and the defining formula is

$$\begin{aligned}(\mathrm{Op}(f)u)(s, x) &= \int f\left((st)^{1/2}, \frac{x+y}{2}; \tau, \eta\right) \exp 2i\pi\ [(s-t)\tau + \langle x-y, \eta\rangle]\ u(t, y)\\ &\quad \left(\frac{s}{t}\right)^{1/2} dt\ dy\ d\tau\ d\eta.\end{aligned} \tag{13.29}$$

If one is interested in functions on $\mathbb{R}^{n+1}_+$ that have traces on the boundary, it is more natural to use $L^2(\mathbb{R}^{n+1}_+; ds\ dx)$ (and the related Sobolev spaces) as basic Hilbert spaces. We thus introduce

$$Q(f) = (s^{-1/2})\mathrm{Op}(f)(s^{1/2}), \tag{13.30}$$

where (s^α) stands for the operator of multiplication by s^α, and note that the formula which defines $(Q(f)u)(s,x)$ is just the same as (13.29), except that the factor $(s/t)^{1/2}$ on the right-hand side has to be deleted.

Some readers will want to know what is the relationship between the calculus $f \mapsto Q(f)$ and Melrose's *totally characteristic* calculus: his is a calculus of operators on $\mathbb{R}_+^{n+1}$ characterized by the defining formula (none other than a restriction of the standard calculus on $\mathbb{R}^{n+1}$)

$$(\mathrm{Op}_0(\tilde{a})u)(s, x) = \int \tilde{a}(s, x; \sigma, \xi) e^{2i\pi(s\sigma + \langle x,\xi\rangle)} \hat{u}(\sigma, \xi) d\sigma\, d\xi, \qquad (13.31)$$

where $\tilde{a}$ is linked to some classical symbol a of order k in the sense of the standard (or Weyl) calculus on $\mathbb{R}^{n+1}$ by the formula

$$\tilde{a}(s, x; \sigma, \xi) = a(s, x; s\sigma, \xi). \qquad (13.32)$$

In a sense which was already alluded to above for the half-line case, the totally characteristic calculus should be regarded as a calculus of "functions" of the operators s, x_j, $(1/2i\pi)\, s\, d/ds$, $(1/2i\pi)\, \partial/\partial x_k$ ($j,k = 1, \ldots, n$): the degeneracy in the normal derivative is imposed by (13.32). Even though one is only interested in the values of $(\mathrm{Op}_0(\tilde{a})\, u)\, (s,x)$ for $s \geq 0$, the right-hand side of (13.31), say in the case when $u \in \mathcal{S}(\mathbb{R}^{n+1})$, will generally not depend on the values of $u(t,y)$ for $t \geq 0$ only: to force this condition, Melrose had to assume that the symbol a is *lacunary,* i.e., satisfies

$$\int a(s, x; \sigma, \xi)\, e^{2i\pi r\sigma} d\sigma = 0 \quad \text{for} \quad r \geq 1. \qquad (13.33)$$

For a reason that was already pointed to in the beginning of the introduction, such an assumption is not necessary if, instead of Op_0, one uses the $f \mapsto Q(f)$ calculus. So as to concentrate to what happens near $s = 0$, not at infinity, introduce, with Hörmander [H5; p. 113] the class S_+^k of symbols on $\mathbb{R}^{n+1}$ characterized by the set of inequalities

$$|\partial_s^p \partial_x^\alpha \partial_\sigma^q \partial_\xi^\beta a(s, x; \sigma, \xi)| \leq C(1 + s)^{-N}(1 + |\sigma| + |\xi|)^{k-q-|\beta|} \qquad (13.34)$$

with N arbitrarily large and $C = C(p,\alpha,q,\beta,N)$. We are then in a position to quote Theorem 16.2 in [U6]: *the class of operators* $Q(\tilde{a})$ *with* $a \in S_+^k$ *coincides with that of operators* $\mathrm{Op}_0(\tilde{b})$ *where* $b \in S_+^k$ *is lacunary.*

An instructive example is provided by the half-cylinder $M = \{x = (x_0, x_*) \in \mathbb{R} \times \mathbb{R}^n : x_0 > 0, |x_*| < 1\}$. If one looks at it as a product $\mathbb{R}_+ \times B_n$, totally characteristic differential operators will appear as built from the following basic first-order operators:

$$x_0 \frac{\partial}{\partial x_0}, \qquad \sum_1^n \alpha_j \frac{\partial}{\partial x_j} \text{ with } |\alpha| = 1 \text{ and } \langle \alpha, x_* \rangle = 0,$$

finally the operator $(1 - |x_*|^2) \Sigma x_j \partial/\partial x_j$; thus, near the boundary of B_n, only the normal derivative has to be replaced by some degenerate version. There exists, however, a symbolic calculus of operators on M more degenerate at the boundary, in which the basic first-order differential operators are $x_0 \partial/\partial x_0$ and the operators $(1 - |x_*|^2) \partial/\partial x_j$ for $1 \le j \le n$. We shall be brief at this point: details are provided in [U6; pp. 143–150]. The idea is to start from the light-cone Λ in $\mathbb{R}^{1+n}$, given as the set of $t = (t_0, t_*)$ with $t_0 > |t_*|$, and to perform the diffeomorphism $\Phi : M \to \Lambda$ defined by

$$t_0 = x_0(1 + |x_*|^2), \qquad t_* = 2x_0 x_*. \tag{13.35}$$

From the structure of symmetric cone of Λ, M inherits the Riemannian structure defined by

$$ds^2 = \left[\frac{dx_0}{x_0} - \frac{2\langle x_*, dx_* \rangle}{1 - |x_*|^2}\right]^2 + \frac{4|dx_*|^2}{(1 - |x_*|^2)^2}, \tag{13.36}$$

which hints at once towards the appearance of the first-order differential operators, degenerate at ∂M, listed above. One can of course transfer the Fuchs calculus from Λ to M, using in particular the cotangent map $T^*\Phi$ to transfer symbols from $T^*\Lambda$ to T^*M. The euclidean structure $\xi \mapsto |\xi|_x$ on $T_x^*(M)$, identified with $\mathbb{R}^{n+1}$ in the canonical way through the inclusion map $M \subset \mathbb{R}^{n+1}$, turns out to be equivalent, in a uniform way, with the one that would be defined by

$$|\xi|_x^2 = x_0^2\, \xi_0^2 + (1 - |x_*|^2)^2 |\xi_*|^2 \tag{13.37}$$

with $\xi = (\xi_0, \xi_*)$. Under the transfer, the class of classical symbols of order k on Λ becomes the class of C^∞ symbols $f = f(x,\xi)$ on $M \times \mathbb{R}^{n+1}$ characterized by the set of inequalities

$$\left|\left(\frac{\partial}{\partial x_0}\right)^p \left(\frac{\partial}{\partial x_*}\right)^\alpha \left(\frac{\partial}{\partial \xi_0}\right)^q \left(\frac{\partial}{\partial \xi_*}\right)^\beta f(x, \xi)\right| \le C x_0^{q-p} (1 - |x_*|^2)^{|\beta| - |\alpha|} (1 + |\xi|_x)^{k-q-|\beta|}. \tag{13.38}$$

If $k \in \mathbb{N}$, differential operators in this class are just the differential operators of order k on M which belong to the algebra generated by the operators $x_0 \partial/\partial x_0$, $(1 - |x_*|^2) \partial/\partial x_j$ and by the operators of multiplication by smooth functions a on M satisfying the estimates

$$\left|\left(\frac{\partial}{\partial x_0}\right)^p \left(\frac{\partial}{\partial x_*}\right)^\alpha a(x)\right| \leq C\, x_0^{-p}\, (1 - |x_*|^2)^{-|\alpha|}. \tag{13.39}$$

As a last example, consider the ball B_n itself. The usual totally characteristic calculus would yield algebras of operators modeled on first-order differential operators among which only the operator of normal differentiation would degenerate like $(1 - |x|^2) \Sigma x_j\, \partial/\partial x_j$. One might surmise from the half-cylinder model, on the other hand, that one should be able to build on B_n a pseudodifferential analysis for which, in a sense that should be clear by now, the basic first-order differential operators would be the operators $(1 - |x|^2)\, \partial/\partial x_j$ $(1 \leq j \leq n)$. This is indeed the case, and the algebra is described in [U7; pp. 164–167]. It is a by-product of the Klein-Gordon analysis, which is best introduced (see the Introduction) with some physical motivation. However, in the present context, it may be useful to note that the Klein-Gordon analysis itself may be thought of as a pseudodifferential analysis on the mass hyperboloid

$$\mathcal{M} = \{(t_0, t_*) \in \mathbb{R}^{n+1}\colon t_0 = (1 + |t_*|^2)^{1/2}\}:$$

as such, it is a descendant of the Fuchs analysis on the light-cone Λ which is foliated by dilations of $\mathcal{M}$. The transfer from $\mathcal{M}$ to B_n is realized by means of the diffeomorphism, standard in special relativity, that associates a velocity with the energy-momentum covector of a particle with a given mass.

14

The λ-Weyl Calculus: Unbounded Realization

The Fuchs calculus on the cotangent bundle $T^*\Lambda$ over a symmetric cone Λ can be regarded as a limit, as $\Lambda \to \infty$, of a family of calculi, called the λ-Weyl calculi. These symbolic calculi have a different covariance group related to multivariable complex-analysis. In the following sections we outline the construction of the λ-Weyl calculus and perform the limit, at least on the representation theoretic level.

Let Λ be an *irreducible* symmetric cone of rank r in $\mathbb{R}^n$, and let

$$\Pi := \Lambda + i\mathbb{R}^n$$

be the corresponding half-space, which is an unbounded tube domain in $\mathbb{C}^n$ and an irreducible hermitian symmetric space of rank r (cf. [L2, UU]). Let

$$\mathrm{Aut}(\Pi) := \{\gamma : \Pi \to \Pi : \gamma \text{ biholomorphic}\}$$

be the group of all biholomorphic automorphisms of Π. Since Π is equivalent to a bounded domain via a generalized Cayley transform [U13, L2], H. Cartan's Theorem (cf. [N1]) shows that $\mathrm{Aut}(\Pi)$ is a real Lie group of finite dimension. Let

$$G(\Pi) := \mathrm{Aut}(\Pi)^\circ$$

denote its identity component. By [H7], the group $G(\Pi)$ is generated by the affine transformations

$$P(x + i\xi) := Px + iP\xi \qquad (P \in GL(\Lambda)^\circ),$$

$$T_b(x + i\xi) := x + i(\xi + b) \qquad (b \in \mathbb{R}^n)$$

and the symmetry

$$S(z) := z^{-1}$$

around $e \in \Pi$. In particular, we obtain the *quasi-translations*

$$T^b(z) := (ST_bS)(z) = (z^{-1} + ib)^{-1} \tag{14.1}$$

for $b \in \mathbb{R}^n$.

The λ-Weyl calculus is related to the *holomorphic discrete series* representations of $G(\Pi)$, realized on weighted Bergman spaces of holomorphic functions.

DEFINITION 14.2 The *characteristic multiplicity* of Λ is the integer a defined via the equation

$$n = r + \frac{a}{2} r(r - 1). \tag{14.3}$$

The *genus* p is defined by

$$p := 2 + a(r - 1) = 2\frac{n}{r}. \tag{14.4}$$

Example 14.5

For the matrix cones $\Lambda = \Lambda_{\mathbb{K}}$ over $\mathbb{K} \in \{\mathbb{R}, \mathbb{C}, \mathbb{H}\}$, defined in Examples 2.14 and 2.21, the characteristic multiplicity is

$$a = \dim_{\mathbb{R}}(\mathbb{K}) \in \{1,2,4\}.$$

The light-cone $\Lambda = \Lambda_n$ (cf. Example 2.27) of dimension $n \geq 3$ has characteristic multiplicity $a = n - 2$.

The following result is due to S. Gindikin [G1]*:*

PROPOSITION 14.6

Let p be the genus of Λ. For $\lambda > p - 1$, the Γ-function of Λ (cf. (8.9)) *has the value*

$$\Gamma_\Lambda(\lambda) = (2\pi)^{(n-r)/2} \prod_{j=1}^{r} \Gamma\left(\lambda - \frac{a}{2}(j-1)\right), \tag{14.7}$$

where Γ is Euler's Γ-function.

In order to define the λ-Weyl calculus, let

$$dV(z) = d(\mathrm{Re}\, z)\, d(\mathrm{Im}\, z)$$

be the Lebesgue measure on Z associated with its hermitian structure and consider the measure

$$d\mu_\Pi(z) := \pi^{-n} \frac{\Gamma_\Lambda(p)}{\Gamma_\Lambda(p/2)} dV(z). \tag{14.8}$$

The normalizing factor in (14.8) is related to the euclidean volume of the *bounded* realization of Π (cf. (16.16)). Let

$$L^2(\Pi) := L^2(\Pi, d\mu_\Pi) \tag{14.9}$$

be the corresponding L^2-space of square-integrable functions, with inner product

$$(\varphi|\psi)_\Pi := \int_\Pi \overline{\varphi(z)}\psi(z)d\mu_\Pi(z). \tag{14.10}$$

DEFINITION 14.11 The subspace

$$H^2(\Pi) := \{f \in L^2(\Pi) : f \text{ holomorphic}\}$$

of $L^2(\Pi)$ is called the *Bergman space* over Π.

One can show [H1] that $H^2(\Pi)$ is a closed subspace of $L^2(\Pi)$ and is therefore a Hilbert space with inner product $(\varphi|\psi)_\Pi$ given by (14.10). The orthogonal projection

$$E^\Pi : L^2(\Pi) \to H^2(\Pi) \tag{14.12}$$

is called the *Bergman projection* over Π. Let $\Delta : \mathbb{C}^n \to \mathbb{C}$ be the Jordan algebraic *determinant* function (obtained from the determinant function $\Delta : \mathbb{R}^n \to \mathbb{R}$ by complex-analytic extension). Then we have the following well-known result:

PROPOSITION 14.13
The reproducing "Bergman" kernel of $H^2(\Pi)$ *is given by*

$$E^\Pi(z, w) = \Delta(z + w^*)^{-p} \tag{14.14}$$

for all $z,w \in \Pi$. *Here* Δ *is the Jordan algebra determinant and* p *is the genus. Thus for every* $w \in \Pi$, *the holomorphic function*

$$E^\Pi_w(z) := E^\Pi(z, w) \tag{14.15}$$

on Π *belongs to* $H^2(\Pi)$, *and the Bergman projection (14.12) has the form*

$$(E^\Pi \varphi)(z) = (E^\Pi_z | \varphi)_\Pi \tag{14.16}$$

for all $\varphi \in L^2(\Pi)$ *and* $z \in \Pi$.

PROOF By definition of the Bergman space, we have

$$\text{Det}\, \partial\gamma(z) \cdot E^\Pi(\gamma(z), \gamma(w)) \cdot \overline{\text{Det}\, \partial\gamma(w)} = E^\Pi(z, w) \tag{14.17}$$

for all $z,w \in \Pi$ and all biholomorphic automorphisms γ of Π. Here $\partial\gamma(z) \in GL(n,\mathbb{C})$ is the complex derivative. For $z = x + i\xi \in \Pi$, the automorphism

$$\gamma := T_\xi \circ P_x^{1/2} \in \text{Aut}(\Pi) \tag{14.18}$$

[cf. (2.105) and (2.106)] satisfies $\gamma(e) = z$ and

$$\text{Det}\, \partial\gamma(e) = \text{Det}\, \partial T_\xi(x) \text{Det}\, P_x^{1/2} = \Delta(x)^{p/2}.$$

Applying (14.17) to $z = w = e$ we obtain

$$E^\Pi(z, z) = 2^{-2n} \Delta\left(\frac{z + z^*}{2}\right)^{-p} = \Delta(z + z^*)^{-p}.$$

Since (14.14) is sesqui-holomorphic in (z,w), the assertion follows.

Q.E.D.

By (14.17), the measure

$$d\mu(z) := E^{\Pi}(z, z) \cdot dV(z) = \Delta(z + z^*)^{-p} dV(z)$$

is invariant under Aut(Π). This measure does not give rise to an interesting Hilbert space of analytic functions. Instead, let $\lambda > p - 1$ and consider the measure

$$d\mu_\lambda(z) := \pi^{-n} \frac{\Gamma_\Lambda(\lambda)}{\Gamma_\Lambda\left(\lambda - \frac{p}{2}\right)} \Delta(z + z^*)^{\lambda - p} dV(z), \tag{14.19}$$

with normalizing constant involving the Γ-function (14.7). Consider the L^2-space

$$L^2_\lambda(\Pi) := L^2(\Pi, d\mu_\lambda) \tag{14.20}$$

of square-integrable functions on Π, with inner product

$$(\varphi \mid \psi)_\lambda := \int_\Pi \overline{\varphi(z)} \psi(z) d\mu_\lambda(z). \tag{14.21}$$

For $\lambda = p$, we obtain the space $L^2_p(\Pi) = L^2(\Pi)$ defined in (14.9).

DEFINITION 14.22 The subspace

$$H^2_\lambda(\Pi) := \{f \in L^2_\lambda(\Pi) : f \text{ holomorphic}\}$$

of $L^2_\lambda(\Pi)$ is called the λ-*Bergman space* over Π.

One can show that $H^2_\lambda(\Pi)$ is a closed subspace of $L^2_\lambda(\Pi)$ and is therefore a Hilbert space with inner product $(\varphi|\psi)_\lambda$ given by (14.21). The orthogonal projection

$$E^\lambda : L^2_\lambda(\Pi) \to H^2_\lambda(\Pi) \tag{14.23}$$

is called the λ-*Bergman projection* over Π. In the special case $\lambda = p$, we obtain the standard Bergman space

$$H^2_p(\Pi) = H^2(\Pi)$$

and the standard Bergman projection $E^p = E^{\Pi}$.

PROPOSITION 14.24
The reproducing "λ-Bergman" kernel of $H^2_\lambda(\Pi)$ *is given by*

$$E^\lambda(z, w) = \Delta(z + w^*)^{-\lambda} \tag{14.25}$$

for all $z,w \in \Pi$. *Here* Δ *is the Jordan algebra determinant,* $\lambda > p - 1$ *and the holomorphic branch* $z \mapsto \Delta(z)^{-\lambda}$ *on the convex domain* Π *is determined by* $\Delta(e)^{-\lambda} := 1$.

Thus for every $w \in \Pi$, the holomorphic function

$$E^\lambda_w(z) := E^\lambda(z, w) \tag{14.26}$$

on Π belongs to $H^2_\lambda(\Pi)$, and the λ-Bergman projection (14.23) has the form

$$(E^\lambda \varphi)(z) = (E^\lambda_z | \varphi)_\lambda \tag{14.27}$$

for all $\varphi \in L^2_\lambda(\Pi)$ and $z \in \Pi$. For $\lambda > p - 1$, the weighted Bergman space $H^2_\lambda(\Pi)$ carries a unitary projective representation

$$U^\Pi_\lambda : G(\Pi) \to U(H^2_\lambda(\Pi)) \qquad \text{(unitary group of } H^2_\lambda(\Pi)) \tag{14.28}$$

defined as follows: for every $\gamma \in G(\Pi)$ choose a holomorphic branch $z \mapsto [\text{Det } \partial\gamma(z)]^{\lambda/p}$ on Π and put

$$(U^\Pi_\lambda(\gamma^{-1})f)(z) := f(\gamma z)[\text{Det } \partial\gamma(z)]^{\lambda/p} \tag{14.29}$$

for all $f \in H^2_\lambda(\Pi)$ and $z \in \Pi$. Then the "phase factors" $U^\Pi_\lambda(\gamma_1\gamma_2)^{-1}$ $U^\Pi_\lambda(\gamma_1)\, U^\Pi_\lambda(\gamma_2)$, for $\gamma_1,\gamma_2 \in G(\Pi)$, belong to the group $\exp(2\pi i\, \lambda/p\, \mathbb{Z})$. In particular, (14.29) defines a true (not only a projective) representation of $G(\Pi)$ if λ is an integer multiple of p. For the symmetry $S(z) = S_e(z) = z^{-1}$, belonging to $G(\Pi)$, one has

$$(U^\Pi_\lambda(S_e)f)(z) = f(z^{-1})\Delta(z)^{-\lambda} e^{i\pi n(\lambda/p)}, \tag{14.30}$$

say, with $\Delta(z)^{-\lambda}$ normalized by $\Delta(e)^{-\lambda} = 1$ (cf. Proposition 14.24), since (3.9) implies

$$\text{Det } \partial S_e(z) = \text{Det}(-P_z^{-1}) = (-1)^n \Delta(z)^{-p}.$$

Define

$$\mathrm{Sym}_e^{\lambda}(f)(z) := f(z^{-1})\Delta(z)^{-\lambda} \tag{14.31}$$

so as to get a self-adjoint operator (which is both unitary and involutive). For the symmetries $S_{x+i\xi}$ about any point $x + i\xi \in \Pi$ we use the affine automorphism γ defined in (14.18) and put

$$\mathrm{Sym}_{x+i\xi}^{\lambda} = e^{-i\pi n(\lambda/p)} U_{\lambda}^{\Pi}(S_{x+i\xi}), \tag{14.32}$$

where the determination of the phase factor $e^{-i\pi n(\lambda/p)}$ is that which makes $\mathrm{Sym}_{x+i\xi}^{\lambda}$ conjugate to the operator Sym_e^{λ} under some element in $U(H_{\lambda}^2(\Pi))$.

DEFINITION 14.33 Let $f : \Pi \to \mathbb{C}$ be summable for the invariant measure $d\mu$. For each $\lambda > p - 1$, the λ-*Weyl operator* with *active symbol* f is given by the operator integral

$$\sigma_{\lambda}^{*}(f) = 2^n \int_{\Pi} f(x + i\xi)\mathrm{Sym}_{x+i\xi}^{\lambda} d\mu(x + i\xi), \tag{14.34}$$

acting on the weighted Bergman space $H_{\lambda}^2(\Pi)$. Conversely, for any trace-class operator A on $H_{\lambda}^2(\Pi)$, the smooth function

$$\sigma_{\lambda}(A)(x + i\xi) := 2^n \,\mathrm{trace}(A \cdot U_{\lambda}(s_{x+i\xi})) \tag{14.35}$$

on Π is the *passive* λ-*Weyl symbol* of A. The two maps σ_{λ}^{*} and σ_{λ} are formally adjoint to each other when viewed as maps between the space of square-summable functions on Π (with respect to $d\mu$) and the space of Hilbert-Schmidt operators on $H_{\lambda}^2(\Pi)$, as follows from the definition and the fact that the operators Sym_z^{λ} themselves are self-adjoint.

PROPOSITION 14.36
The λ-Weyl calculus has the covariance group $G(\Pi) = \mathrm{Aut}(\Pi)^{\circ}$. *This means that*

$$\sigma_{\lambda}^{*}(f \circ \gamma^{-1}) = U_{\lambda}(\gamma)\sigma_{\lambda}^{*}(f)U_{\lambda}(\gamma)^{-1} \tag{14.37}$$

for all integrable symbols f *and all transformations* $\gamma \in G(\Pi)$, *and a similar relation holds for the passive symbol.*

PROOF Use the fact that the symmetries s_{ζ} for $\zeta \in \Pi$ satisfy $\gamma s_{\zeta}\gamma^{-1} = s_{\gamma(\zeta)}$ for all $\gamma \in G(\Pi)$. Q.E.D.

15

Contraction of the λ-Weyl Calculus

The λ-Weyl calculus over the tube domain $\Pi = \Lambda + i\mathbb{R}^n$ associated with a symmetric cone Λ contracts into the Fuchs calculus over Λ, as the parameter $\lambda \to \infty$. The Fuchs calculus is a "partially flat" limit of the λ-Weyl calculus (the geometry of the base Λ remains unchanged). In this section we describe the asymptotic behavior, as $\lambda \to \infty$, of the representation U_λ^Π of $G(\Pi)$, defined in (14.29), which underlies the λ-Weyl calculus (14.34). We also need to deform the geometric action of $G(\Pi)$ on the underlying phase space Π (which preserves the Kähler structure and hence the symplectic structure of Π). This is because the covariance group Γ of the limiting Fuchs calculus arises as a contraction of $G(\Pi)$. We will first study the limit $\lambda \to \infty$ on the classical level. Suppose in the following that $\Lambda \subset \mathbb{R}^n$ is an irreducible symmetric cone, with linear automorphism group $GL(\Lambda)$. Let

$$T^*\Lambda = \Lambda \times \mathbb{R}^n$$

denote the *cotangent bundle* of Λ, endowed with its canonical symplectic structure. The group

$$\mathrm{Aut}(T^*\Lambda)$$

of all *symplectomorphisms* of $T^*\Lambda$ is an infinite dimensional "Lie group". Now consider the diffeomorphism $\varphi_\lambda : \Pi \to T^*\Lambda$ defined by

$$\varphi_\lambda\,(x + i\xi) := (x, -\lambda P_x^{-1}\xi). \tag{15.1}$$

It is essential to regard Π as $T\Lambda$ (not $T^*\Lambda$) with a complex structure. Then φ_λ is linear in the fibers of $T\Lambda$ and $T^*\Lambda$, respectively, and moreover commutes with $GL(\Lambda)$ if P acts on Π as $P(x + i\xi) = Px + iP\xi$, on $T^*\Lambda$ as

$P(y,\eta) = (Py, P'^{-1}\eta)$. Thanks to φ_λ, $T^*\Lambda$ is given a Riemannian structure, euclidean in the fibers. For every biholomorphic mapping $\gamma \in \mathrm{Aut}(\Pi)$,

$$\mathrm{Ad}(\varphi_\lambda)\gamma := \varphi_\lambda \circ \gamma \circ \varphi_\lambda^{-1} \tag{15.2}$$

defines a diffeomorphism of $T^*\Lambda$.

PROPOSITION 15.3

The generators of the Fuchs covariance group $\Gamma \subset \mathrm{Aut}(T^*\Lambda)$ *(cf. Proposition 3.3) are given as follows:*

$$\tilde{P}(x, \xi) := \mathrm{Ad}(\varphi_\lambda)(P)(x, \xi) = (Px, P'^{-1}\xi) \qquad (P \in GL(\Lambda)), \tag{15.4}$$

$$\tilde{S}(x, \xi) := \lim_{\lambda\to\infty} \mathrm{Ad}(\varphi_\lambda)(S)(x, \xi) = (x^{-1}, -P_x\xi), \tag{15.5}$$

$$\tilde{T}^b(x, \xi) := \mathrm{Ad}(\varphi_\lambda)(T_{b/\lambda})(x, \xi) = (x, \xi - P_x^{-1}b) \qquad (b \in \mathbb{R}^n), \tag{15.6}$$

$$\tilde{T}_b(x, \xi) := \lim_{\lambda\to\infty} \mathrm{Ad}(\varphi_\lambda)(T^{b/\lambda})(x, \xi) = (x, \xi + b) \qquad (b \in \mathbb{R}^n). \tag{15.7}$$

PROOF Every $P \in GL(\Lambda)$ satisfies

$$PP_xP' = P_{Px}$$

and therefore we obtain for all $(x,\xi) \in T^*\Lambda$:

$$\mathrm{Ad}(\varphi_\lambda)(P)(x, \xi) = \varphi_\lambda\left(Px - \frac{i}{\lambda} PP_x\xi\right) = (Px, P_{Px}^{-1}PP_x\xi) = (Px, P'^{-1}\xi), \tag{15.8}$$

an action independent of λ. Similarly, the translations give

$$\mathrm{Ad}(\varphi_\lambda)(T_{b/\lambda})(x, \xi) = \varphi_\lambda\left(x - \frac{i}{\lambda} P_x\xi + \frac{i}{\lambda} b\right) = (x, \xi - P_x^{-1}b) \tag{15.9}$$

independent of λ. This proves (15.4) and (15.6). On the other hand, $\mathrm{Ad}(\varphi_\lambda)(S)\ (x,\xi) = \varphi_\lambda((x - i/\lambda\ P_x\xi)^{-1})$. Since the derivative of S at x is $-P_x^{-1}$, one has

$$\left(x - \frac{i}{\lambda} P_x \xi\right)^{-1} = x^{-1} + \frac{i}{\lambda} \xi + O(\lambda^{-2}), \qquad \lambda \to \infty$$

so that (15.5) is immediate. Finally, as $P_{x^{-1}} = P_x^{-1}$ for all $x \in \Lambda$, we have

$$\begin{aligned}(\tilde{S}\tilde{T}_b \tilde{S})(x, \xi) &= \tilde{S}(x^{-1}, b - P_x\xi) \\ &= (x, -P_x^{-1}(b - P_x\xi)) = (x, \xi - P_x^{-1}b) = \tilde{T}^b(x, \xi).\end{aligned}$$

Since $T^{b/\lambda} = ST_{b/\lambda}S$ according to (14.1), we obtain from (15.5) and (15.9)

$$\lim_{\lambda\to\infty} \mathrm{Ad}(\varphi_\lambda)(T^{b/\lambda}) = \tilde{S}\tilde{T}^b\tilde{S} = \tilde{T}_b. \qquad \text{Q.E.D.}$$

The limiting process described in Proposition 15.3 is a Lie algebra contraction, a construction which, in the particular case at hand, may be described as follows. Suppose a Lie algebra $\mathfrak{g}$ has a splitting

$$\mathfrak{g} = \mathfrak{g}_+ \oplus \mathfrak{g}_-$$

into a subalgebra $\mathfrak{g}_+$ and a subspace $\mathfrak{g}_-$. Let $p_- : \mathfrak{g} \to \mathfrak{g}_-$ be the projection onto $\mathfrak{g}_-$ with null-space $\mathfrak{g}_+$. Define

$$[A_+ + A_-, B_+ + B_-]_c := [A_+, B_+] + p_-([A_+, B_-] + [A_-, B_+]) \qquad (15.10)$$

whenever $A_+, B_+ \in \mathfrak{g}_+$ and $A_-, B_- \in \mathfrak{g}_-$. Denote by $\mathfrak{g}_c$ the vector space $\mathfrak{g}$ endowed with the new product (15.10). Then $\mathfrak{g}_c$ is a Lie algebra which may be called a *contraction* of $\mathfrak{g}$ in so far as some of the Lie structure constants have been replaced by zero. Note that $\mathfrak{g}_+$ is a subalgebra of $\mathfrak{g}_c$ whereas $\mathfrak{g}_-$ is an *abelian ideal* in $\mathfrak{g}_c$. A Lie group G_c is called a *contraction* of a Lie group G if the Lie algebra $\mathfrak{g}_c$ of G_c is a contraction of the Lie algebra $\mathfrak{g}$ of G.

PROPOSITION 15.11
The Fuchs covariance group Γ *is a contraction of* $G(\Pi) := \mathrm{Aut}(\Pi)^\circ$.

PROOF The Lie algebra of Γ (or its identity component Γ°), denoted by Lie(Γ), can be determined by differentiating a smooth curve

$$\theta \mapsto [P_\theta, b_\theta, c_\theta]$$

of transformations (3.11) passing through the unit element [id,0,0]. By (3.11), the Lie algebra of Γ° is generated by the vector fields

$$(\tilde{M} := (Mx, -M'\xi) \qquad (M \in \mathfrak{gl}(\Lambda)), \tag{15.12}$$

$$\tilde{A}_b := (0, b) \qquad (b \in \mathbb{R}^n), \tag{15.13}$$

$$\tilde{A}^c := (0, -P_x^{-1}c) \qquad (c \in \mathbb{R}^n) \tag{15.14}$$

on $T^*\Lambda = \Lambda \times \mathbb{R}^n$. Here

$$\mathfrak{gl}(\Lambda) := \{M \in \mathfrak{gl}(n, \mathbb{R}): \exp(\theta M)\Lambda = \Lambda, \forall\theta \in \mathbb{R}\}$$

is the Lie algebra of $GL(\Lambda)$. Let

$$\{sxt\} := P(s, t)x \qquad (s, t, x \in \mathbb{R}^n) \tag{15.15}$$

be the *Jordan triple product* on $\mathbb{R}^n$, where $P(s,t)$ is defined by (5.6). Then the commutator

$$[(f_1, g_1), (f_2, g_2)] := \left(\frac{\partial f_1}{\partial x} f_2 + \frac{\partial f_1}{\partial \xi} g_2 - \frac{\partial f_2}{\partial x} f_1 - \frac{\partial f_2}{\partial \xi} g_1, \frac{\partial g_1}{\partial x} f_2 + \frac{\partial g_1}{\partial \xi} g_2 - \frac{\partial g_2}{\partial x} f_1 - \frac{\partial g_2}{\partial \xi} g_1\right)$$

of vector fields on $T^*\Lambda$ satisfies

$$[\tilde{A}_b, \tilde{M}] = \tilde{A}_{M'b}, [\tilde{M}_1, \tilde{M}_2] = (M_1M_2 - M_2M_1)^{\sim}$$

and

$$[\tilde{M}, \tilde{A}^c] = (0, M'P_x^{-1}c) - 2\{P_x^{-1}(Mx)cx^{-1}\}) = \tilde{A}^{Mc}$$

for all $M, M_1, M_2 \in \mathfrak{gl}(\Lambda)$ and $b, c \in \mathbb{R}^n$. Here the last identity follows from the formulas

$$M'(P_x^{-1}c) = 2\{(P_x^{-1}Mx)cx^{-1}\} - P_x^{-1}Mc \tag{15.16}$$

and

$$M'(x^{-1}) = P_x^{-1}(Mx), \tag{15.17}$$

valid for all $M \in \mathfrak{gl}(\Lambda)$. These formulas can be obtained by taking the derivative in the corresponding formulas (2.9) and (2.12) for $GL(\Lambda)$. Trivially, we have

$$[\tilde{A}_b, \tilde{A}_c] = [\tilde{A}^b, \tilde{A}_c] = [\tilde{A}^b, \tilde{A}^c] = 0$$

whenever $b,c \in \mathbb{R}^n$. This describes the Lie algebra Lie(Γ). Now consider the complex domain Π. By Cartan's Theorem, the Lie algebra $\mathfrak{g} := \mathrm{aut}(\Pi)$ of the Lie group G(Π) consists of all completely integrable holomorphic vector fields

$$A = h(z)\frac{\partial}{\partial z}$$

on Π. Here $z = (z_1, \ldots, z_n)$ are complex coordinates and $\partial/\partial z$ is a column vector, acting on holomorphic functions only (thus we omit the anti-holomorphic Wirtinger derivatives). Furthermore, $h : \Pi \to \mathbb{C}^n$ is a holomorphic mapping. Complete integrability of A means that there exists a continuous 1-parameter group $\gamma_\theta = \exp(\theta A) \in G(\Pi)$ satisfying the ordinary differential equation

$$\frac{\partial \gamma_\theta(z)}{\partial \theta} = h(\gamma_\theta(z))$$

for all $\theta \in \mathbb{R}$ and $z \in \Pi$. The holomorphic vector fields

$$\hat{M} := Mz\frac{\partial}{\partial z} \qquad (M \in \mathfrak{gl}(\Lambda)), \tag{15.18}$$

$$A_b := ib\frac{\partial}{\partial z} \qquad (b \in \mathbb{R}^n) \tag{15.19}$$

on Π satisfy

$$\exp(\hat{M}) = \exp(M),$$
$$\exp(A_b) = T_b.$$

Here $M \in \mathfrak{gl}(\Lambda) \subset \mathfrak{gl}(n,\mathbb{R})$ is extended to Π by complex linearity. It follows that these vector fields are completely integrable on Π. We put

$$\mathrm{aut}_1(\Pi) = \{A_b : b \in \mathbb{R}^n\}$$

and

$$\mathrm{aut}_0(\Pi) = \{\hat{M} : M \in \mathfrak{gl}(\Lambda)\}.$$

Then

$$\mathrm{aut}_+(\Pi) := \mathrm{aut}_1(\Pi) \oplus \mathrm{aut}_0(\Pi)$$

consists of all *affine* (complete) vector fields on Π. By (14.1), we have for all $b \in \mathbb{R}^n$

$$T^b = ST_bS = \exp(A^b),$$

where

$$A_b = (dS)_* A_b = i\, dS(S^{-1}(z))b \frac{\partial}{\partial z} = -i(P_{z^{-1}})^{-1} b \frac{\partial}{\partial z} = -iP_z b \frac{\partial}{\partial z}. \tag{15.20}$$

We put

$$\mathrm{aut}_{-1}(\Pi) = \{A^b : b \in \mathbb{R}^n\}.$$

Then it is known [U12, L2] that

$$\mathrm{aut}(\Pi) = \mathrm{aut}_+(\Pi) \oplus \mathrm{aut}_{-1}(\Pi) = \mathrm{aut}_1(\Pi) \oplus \mathrm{aut}_0(\Pi) \oplus \mathrm{aut}_{-1}(\Pi). \tag{15.21}$$

Clearly, $\mathrm{aut}_+(\Pi)$ is a subalgebra of $\mathrm{aut}(\Pi)$. The subspace $\mathrm{aut}_{-1}(\Pi)$ is an abelian subalgebra but it is not an ideal in $\mathrm{aut}(\Pi)$. Let $\mathfrak{g}_c$ be the contracted Lie algebra structure on $\mathfrak{g}$ associated with the splitting (15.21). For the commutator

$$\left[h(z) \frac{\partial}{\partial z}, k(z) \frac{\partial}{\partial z}\right] = (\partial h(z) \cdot k(z) - \partial k(z) \cdot h(z)) \frac{\partial}{\partial z} \tag{15.22}$$

of vector fields on Π we obtain

$$[\hat{M}, A_b] = A_{Mb},$$

$$[\hat{M}_1, \hat{M}_2] = (M_1M_2 - M_2M_1)^\wedge$$

and

$$[A^c, \hat{M}] = i(MP_zc - 2\{(Mz), c, z\}) \frac{\partial}{\partial z} = A^{M'c}$$

for all $M, M_1, M_2 \in \mathfrak{gl}(\Lambda)$ and $b, c \in \mathbb{R}^n$. Here we use the complex extension of the Jordan triple product (15.15), and the identity

$$P_z M'c + MP_z c = 2\{(Mz)cz\}$$

which is analogous to (15.16). Since for all b,c $\in \mathbb{R}^n$

$$[A_b, A_c] = 0 = [A^b, A^c]$$

and

$$[A^b, A_c] \in \text{aut}_0(\Pi) \subset \text{aut}_+(\Pi),$$

Lie(Γ) is isomorphic to the contraction $\mathfrak{g}_c$ of $\mathfrak{g} = \text{aut}(\Pi)$ by sending $\tilde{A}_b \mapsto A_b$, $\tilde{M} \mapsto -\hat{M}'$ and $\tilde{A}^c \mapsto A^c$. Q.E.D.

We now pass to the study of the representation U_λ^Π of G(Π), as $\lambda \to \infty$. The first step is a "real" characterization of the weighted Bergman spaces $\overline{H_\lambda^2}(\Pi)$, via a known Paley-Wiener argument. We will actually work with the complex-conjugate space $H_\lambda^2(\Pi)$ of anti-holomorphic functions. Consider the invariant measure dm(t) on Λ (cf. (3.38)) and the associated Hilbert space $L^2(\Lambda)$. Let the invariant inner product $\langle s,t \rangle$ on $\mathbb{R}^n$ be normalized by the condition

$$\langle e,e \rangle = r = \text{rank}(\Lambda)$$

and consider its sesqui-linear extension $(z|w)$ on $\mathbb{C}^n$, linear in the first variable.

THEOREM 15.23
For each $\lambda > p - 1$, *there exists a Hilbert space isomorphism*

$$\overline{H_\lambda^2}(\Pi) \underset{\cong}{\rightarrow} L^2(\Lambda)$$

with adjoint given by the "λ-Laplace transform"

$$(\mathcal{L}_\lambda u)(z) = c_\lambda \cdot \int_\Lambda e^{-2\pi(t|z)} u(t) \Delta(t)^{\lambda - p/2} dt \tag{15.24}$$

for all $u \in L^2(\Lambda)$ *and* $z \in \Pi$. *Here*

$$c_\lambda = \frac{(2\pi)^{\lambda r/2}}{\sqrt{\Gamma_\Lambda(\lambda)}}.$$

PROOF Recalling that $dm(t) = \Delta(t)^{-p/2}\,dt$, one has, according to Definition 8.4,

$$(\mathcal{L}_\lambda u)(z) = c_\lambda \Delta(\text{Re } z)^{-\lambda/2}(\psi_z^\lambda | u)$$

and since, according to Proposition 8.7, one has

$$\int_\Pi |(\psi_z^\lambda | u)|^2 d\mu(z) = (4\pi)^{n-\lambda r}\Gamma_\Lambda\left(\lambda - \frac{p}{2}\right)\|u\|_\Lambda^2,$$

the fact that $\mathcal{L}_\lambda$ is an isometry is an immediate consequence of (14.19). A classical argument (cf. [K2, FK2]) shows that $\mathcal{L}_\lambda$ is surjective. Q.E.D.

DEFINITION 15.25 For $\lambda > p - 1$, define a projective unitary representation U_λ^Λ from $G(\Pi)$ in the Hilbert space $L^2(\Lambda)$ by putting

$$U_\lambda^\Lambda(\gamma) := \mathcal{L}_\lambda^* U_\lambda^\Pi(\gamma)\mathcal{L}_\lambda \tag{15.26}$$

for all $\gamma \in G(\Pi)$. Here $\mathcal{L}_\lambda$ is the λ-Laplace transform and U_λ^Π is defined by (14.28).

PROPOSITION 15.27
When restricted to the subgroup G *of* $G(\Pi)$, *the representation* U_λ^Λ *agrees with* U, *i.e.,*

$$U(\tilde{P}) = U_\lambda^\Lambda(P) \qquad (P \in GL(\Lambda)), \tag{15.28}$$

$$U(\tilde{T}_b) = U_\lambda^\Lambda(T_b) \qquad (b \in \mathbb{R}^n). \tag{15.29}$$

PROOF Let $P \in GL(\Lambda)$ and $z \in \Pi$. Using (6.3) we obtain for all $u \in L^2(\Lambda)$

$$\begin{aligned}
[U_\lambda^\Pi(P)(\mathcal{L}_\lambda u)](z) &= (\mathcal{L}_\lambda u)(P'z)\cdot(\text{Det } P')^{\lambda/p}\\
&= c_\lambda \int_\Lambda e^{-2\pi(t|P'z)}u(t)\ \Delta(t)^{\lambda-p/2}dt\cdot(\text{Det } P')^{\lambda/p}\\
&= c_\lambda \int_\Lambda e^{-2\pi(Pt|z)}u(t)\Delta(t)^{\lambda-p/2}\cdot(\text{Det } P)^{(\lambda/p)-1}(\text{Det } P)dt\\
&= c_\lambda \int_\Lambda e^{-2\pi(s|z)}u(P^{-1}s)\Delta(s)^{\lambda-p/2}ds\\
&= \mathcal{L}_\lambda(u\circ P^{-1})(z)
\end{aligned} \tag{15.30}$$

independent of λ. This proves (15.28). For $b \in \mathbb{R}^n$, (6.3) gives

$$
\begin{aligned}
U_\lambda^\Pi(T_b)(\mathcal{L}_\lambda u)(z) &= (\mathcal{L}_\lambda u)(z + ib) \\
&= c_\lambda \int_\Lambda e^{-2\pi(t|z+ib)} u(t)\Delta(t)^{\lambda-p/2} dt \\
&= c_\lambda \int_\Lambda e^{-2\pi(t|z)} e^{2\pi i(t|b)} u(t)\Delta(t)^{\lambda-p/2} dt \\
&= \mathcal{L}_\lambda(e^{2\pi i(t|b)} \cdot u)(z). \qquad (15.31)
\end{aligned}
$$

This proves (15.29) for the translations. Q.E.D.

We will now pass to the *infinitesimal generators.* For every $A = h(z)\,\partial/\partial z \in \mathfrak{g} := \operatorname{aut}(\Pi)$ consider the infinitesimal generator, in the sense of Stone's theorem, of the unitary group $\theta \mapsto U_\lambda^\Pi(e^{\theta A})$, i.e., the self adjoint operator

$$(2i\pi)^{-1} dU_\lambda^\Pi(A) = (2i\pi)^{-1} \frac{d}{d\theta} (U_\lambda^\Pi(e^{\theta A}))(\theta = 0). \qquad (15.32)$$

Then one has

$$dU_\lambda^\Pi(A)f(z) = -\partial f(z)h(z) - \frac{\lambda}{p} f(z) \cdot \operatorname{trace} \partial h(z) \qquad (15.33)$$

in $H_\lambda^2(\Pi)$, where $\partial f(z) : \mathbb{C}^n \to \mathbb{C}$ and $\partial h(z) : \mathbb{C}^n \to \mathbb{C}^n$ are the complex derivatives. For the proof of (15.33), consider the one-parameter group

$$\varphi_\theta(z) = \varphi(\theta, z) = \exp(\theta A)(z)$$

associated with A, whose complex derivative satisfies

$$\frac{\partial}{\partial \theta} \partial \varphi_\theta(z) = \frac{\partial^2 \varphi(\theta, z)}{\partial \theta\, \partial z} = \frac{\partial^2 \varphi(\theta, z)}{\partial z\, \partial \theta} = \partial h(z)$$

at $\theta = 0$. Since

$$\frac{\partial}{\partial \theta} \operatorname{Det} A_\theta = \operatorname{trace} \frac{\partial A_\theta}{\partial \theta}$$

at $\theta = 0$ for every curve $\theta \mapsto A_\theta \in GL(n,\mathbb{C})$ with $A_0 = \mathrm{id}$, this proves (15.33).

PROPOSITION 15.34
For $A^b := -iP_z\, b\, \partial/\partial z$ ($b \in \mathbb{R}^n$), *we have*

$$dU_\lambda^\Lambda(A^b)u = -\frac{1}{2\pi}\frac{\lambda(\lambda - p)}{4}(x^{-1}|b)u(x) + \frac{i}{2\pi}L_b u, \tag{15.35}$$

where

$$L_b u := \sum_{j,k}(\{b_j b b_k\}|x)\partial_j\partial_k u \tag{15.36}$$

is a second-order differential operator on Λ *involving the standard orthonormal basis* $b_1, \ldots, b_n$ *of* $\mathbb{R}^n$ *and the Jordan triple product (15.15). Thus*

$$\lim_{\lambda\to\infty}\frac{8\pi}{\lambda^2}dU_\lambda^\Lambda(A^b)u = -i(x^{-1}|b)u(x). \tag{15.37}$$

PROOF Since $\{b_j : 1 \le j \le n\}$ is an orthonormal basis of the irreducible Jordan algebra $\mathbb{R}^n$ we have

$$\sum_j(\{xyb_j\}|b_j) = \frac{p}{2}(x|y)$$

for all $x,y \in \mathbb{R}^n$. Hence the derivative $\partial h(z)y = 2i\{zby\}$ of $h(z) = i\{zbz\}$ has the trace

$$\text{trace } \partial h(z) = ip\cdot(z|b). \tag{15.38}$$

Now let $u \in L^2(\Lambda)$, $v(x) := u(x)\,\Delta(x)^{\lambda-p/2}$ and

$$f(z) = \int_\Lambda e^{-2\pi(z|x)}v(x)dx,$$

so that $c_\lambda \cdot \bar{f} = \mathcal{L}_\lambda(\bar{u})$. Then (15.33) and (15.38) imply

$$dU_\lambda^\Pi\left(-iP_z b\frac{\partial}{\partial z}\right)f(z) = i\partial f(z)\{zbz\} + i\lambda f(z)\cdot(z|b).$$

Expanding $\{zbz\}$ we obtain

$$\begin{aligned}\partial f(z)\{zbz\} &= -2\pi \int_\Lambda e^{-2\pi(z|x)}(\{zbz\}|x)v(x)dx \\ &= -2\pi \sum_{j,k} \int_\Lambda e^{-2\pi(z|x)}(z|b_j)(z|b_k)(\{b_jbb_k\}|x)v(x)dx \\ &= \frac{-1}{2\pi} \sum_{j,k} \int_\Lambda (\partial_j\partial_k e^{-2\pi(z|x)})(\{b_jbb_k\}|x)v(x)dx \\ &= \frac{-1}{2\pi} \sum_{j,k} \int_\Lambda e^{-2\pi(z|x)}\partial_j\partial_k(\{b_jbb_k\}|x)v(x)dx.\end{aligned}$$

Putting

$$\text{grad } v := \sum_j \partial_j v \cdot b_j,$$

we have

$$\begin{aligned}f(z)(z|b) &= \frac{-1}{2\pi} \int_\Lambda (\text{grad } e^{-2\pi(z|x)}|b)v(x)dx \\ &= \frac{1}{2\pi} \int_\Lambda e^{-2\pi(z|x)}(\text{grad } v|b)dx.\end{aligned}$$

According to (5.8), we have

$$\text{grad } v = \Delta(x)^{\lambda-p/2}\cdot \text{grad } u + \frac{\lambda-p}{2} x^{-1}\cdot v(x). \tag{15.39}$$

Using the "associativity" $(\{xyz\}|w) = (x|\{yzw\})$ for $x,y,z,w \in \mathbb{R}^n$ this implies

$$\begin{aligned}\sum_k \partial_k(\{b_jbb_k\}|x)\cdot v(x) &= \frac{p}{2}(b_j|b)v(x) + (\{b_jb\cdot \text{grad } v\}|x) \\ &= \frac{p}{2}(b_j|b)v(x) + \Delta(x)^{\lambda-p/2}(\{b_jb\cdot \text{grad } u\}|x) \\ &\quad + \frac{\lambda-p}{2} v(x)(\{b_jbx^{-1}\}|x) \\ &= \frac{\lambda}{2}(b_j|b)v(x) + \Delta(x)^{\lambda-p/2}\cdot(\text{grad } u|\{bb_jx\})\end{aligned}$$

hence

$$\sum_{j,k} \partial_j\partial_k(\{b_j b b_k\} \,|\, x) \cdot v(x)$$

$$= \frac{\lambda}{2}(\text{grad } v \,|\, b) + \frac{\lambda - p}{2} \Delta(x)^{\lambda - p/2}(\text{grad } u \,|\, \{bx^{-1}x\})$$

$$+ \Delta(x)^{\lambda - p/2} L_b u + \frac{p}{2} \Delta(x)^{\lambda - p/2}(\text{grad } u \,|\, b). \tag{15.40}$$

Combining (15.40) with (15.39), we obtain (15.35). Q.E.D.

The stabilizer subgroup $K_\Pi = \{g \in G(\Pi) : g(e) = e\}$ at $e \in \Pi$ has the Lie algebra $\mathfrak{k}_\Pi$ which is generated by all vector fields

$$i(b - P_z b) \frac{\partial}{\partial z}$$

where $b \in \mathbb{R}^n$ [U12, L2]. Applying (15.35), we obtain

$$dU_\lambda^\Lambda\left(i(b - P_z b) \frac{\partial}{\partial z}\right) u = -2\pi i(x \,|\, b)u(x)$$

$$- \frac{i}{2\pi} \frac{\lambda(\lambda - p)}{4} (x^{-1} \,|\, b)u(x) + \frac{i}{2\pi} L_b u \tag{15.41}$$

hence

$$\lim \frac{8\pi}{\lambda^2} dU_\lambda^\Lambda\left(i(b - P_z b) \frac{\partial}{\partial z}\right) u = -i(x^{-1} \,|\, b)u(x). \tag{15.42}$$

REMARK 15.43

Proposition 15.27 and (15.41) show that the λ-dependence of $U_\lambda^\Lambda \approx U_\lambda^\Pi$ is reflected entirely in the restriction of the representation on the special one-parameter subgroup

$$\exp \theta i(e - P_z e) \frac{\partial}{\partial z} \qquad (\theta \in \mathbb{R}) \tag{15.43}$$

of $G(\Pi)$ tied to the complex structure of Π, since (15.44) together with the subgroup of affine transformations (on which the representation is essentially independent of λ) generate the whole of $G(\Pi)$.

16

The λ-Weyl Calculus: Bounded Realization

The (unbounded) tube domain Π which is the classical phase space underlying the λ-Weyl calculus (in its complex, unbounded realization) is holomorphically equivalent to a bounded symmetric domain via a generalized Cayley transformation. This fact generalizes the well-known biholomorphic correspondence between the unit disk and the half-plane in one complex dimension. In the n-dimensional case the unit disk is replaced by a *bounded symmetric domain* Ω in $\mathbb{C}^n$. Using the Jordan algebraic structure on $\mathbb{C}^n$ associated with Λ and Π, one can describe Ω as the open unit ball of $\mathbb{C}^n$ with respect to a suitable norm $\|\cdot\|$ (which agrees with the euclidean norm only when the rank is one).

DEFINITION 16.1 Endow $\mathbb{C}^n$ with the Jordan product (2.95) and the involution (2.92). The $\mathbb{R}$-trilinear composition

$$\{xy^*z\} := x \circ (y^* \circ z) + z \circ (y^* \circ x) - y^* \circ (x \circ z) \tag{16.2}$$

for $x,y,z \in \mathbb{C}^n$ is called the *Jordan triple product.*

Of course one can also define a complex trilinear composition $\{xyz\}$ on $\mathbb{C}^n$, replacing y by y^*. However, it turns out that the triple product (16.2) makes sense beyond the Jordan algebra case, for example, for spaces of non-square matrices. Thus the triple product (16.2) which is anti-linear in y is the more natural one. It is commutative, i.e.

$$\{xy^*z\} = \{zy^*x\}$$

for all $x,y,z \in \mathbb{C}^n$ and satisfies a "Jordan triple identity" which generalizes the Jordan algebra identity (cf. [U12, L2]). Now fix $x,y \in \mathbb{C}^n$. Then

$$(x \Box y^*)z := \{xy^*z\} \tag{16.3}$$

defines a linear endomorphism of $\mathbb{C}^n$, which in view of (16.2) is also given by

$$x \Box y^* = M_x M_{y^*} - M_{y^*} M_x + M_{y^* \circ x}$$

where $M_x z := x \circ z$ denotes the Jordan algebra multiplication operator on $\mathbb{C}^n$. One can show that $z \Box z^*$ has only non-negative eigenvalues for any $z \in \mathbb{C}^n$. Moreover, the maximal eigenvalue

$$\|z\|_\infty^2 := \sup \operatorname{Spec}(z \Box z^*) \tag{16.4}$$

defines a *norm* on the complex vector space $\mathbb{C}^n$.

DEFINITION 16.5 The open unit ball

$$\Omega := \{z \in \mathbb{C}^n : \|z\|_\infty < 1\} \tag{16.6}$$

is called the *symmetric ball* associated with the involutive Jordan algebra structure on $\mathbb{C}^n$.

Example 16.7

*Let $\mathbb{C}^{r\times r}$ be the Jordan *-algebra of all complex $(r\times r)$-matrices, with anticommutator product and the usual matrix involution. Then (16.2) shows that*

$$\{xy^*z\} = \frac{xy^*z + zy^*x}{2} \tag{16.8}$$

for all $x,y,z \in \mathbb{C}^{r\times r}$. It follows that

$$z \Box z^* = \frac{1}{2}(L_{zz^*} + R_{z^*z})$$

where L and R denote the associative left and right multiplication operators, respectively. It is well known that

$$\sup \operatorname{Spec} z \Box z^* = \|z\|_\infty^2,$$

where $\|z\|_\infty$ is the usual operator norm of matrices. Thus

$$\Omega = \{z \in \mathbb{C}^{r\times r} : z^*z < I_r\}$$

becomes the matrix unit ball in the operator norm. Here I_r *is the* $(r\times r)$*-unit matrix. For* $r = 1$, *we obtain the open unit disk in* $\mathbb{C}$.

Let Ω be the symmetric ball associated with an irreducible complex Jordan *-algebra. Let

$$\mathrm{Aut}(\Omega) := \{\gamma : \Omega \to \Omega \,;\, \gamma \quad \text{biholomorphic}\}$$

be the group of all biholomorphic automorphisms of Ω. Since Ω is a bounded domain, it follows from H. Cartan's Theorem [N1] that $\mathrm{Aut}(\Omega)$ is a real Lie group. Let

$$G(\Omega) := \mathrm{Aut}(\Omega)^\circ$$

denote its identity component. The Jordan triple structure gives an explicit description of $G(\Omega)$. Let

$$GL(\Omega) := \{P \in GL(n,\mathbb{C}) : P\Omega = \Omega\} \tag{16.9}$$

denote the group of all linear transformations preserving Ω. Every element $P \in GL(\Omega)$ preserves the Jordan triple product, and every $\gamma \in \mathrm{Aut}(\Omega)$ fixing the origin $0 \in \Omega$ is linear. Thus

$$GL(\Omega) = \{\gamma \in \mathrm{Aut}(\Omega) : \gamma(0) = 0\}$$

coincides with the stabilizer subgroup at 0. The group $GL(\Omega)$ preserves an inner product on $\mathbb{C}^n$ but, if $r > 1$, $GL(\Omega)$ is a proper subgroup of $U(n)$.

In order to describe non-linear automorphisms of Ω, define the *Bergman endomorphism*

$$(B(x,y)z := z - 2\{xy^*z\} + \{x\{yz^*y\}^*x\} \quad (z \in \mathbb{C}^n). \tag{16.10}$$

for any pair $(x,y) \in \mathbb{C}^n \times \mathbb{C}^n$. Note that (16.10) is complex-linear in z, holomorphic in x and anti-holomorphic in y. For any pair $(z,\zeta) \in \Omega \times \Omega$, the endomorphism $B(z,\zeta)$ is invertible and the vector

$$z^\zeta := B(z,\zeta)^{-1}\,(z - \{z\zeta^*z\}) \tag{16.11}$$

is called the *quasi-inverse* of z with respect to ζ. Clearly, (16.11) is holomorphic in z and anti-holomorphic in ζ. Now fix $b \in \mathbb{C}^n$ and write $\zeta = \tanh(b)$ in the Jordan theoretic sense [L2]. By [L2],

$$\tau_b(z) := -\zeta + B(\zeta,\zeta)^{1/2} z^{\zeta} \tag{16.12}$$

defines an automorphism $\tau_b \in G(\Omega)$ satisfying $\tau_b(0) = -\zeta$ and $\tau_b(\zeta) = 0$. It follows that $G(\Omega)$ acts transitively on Ω. Every $\gamma \in \mathrm{Aut}(\Omega)$ has a unique representation

$$\gamma = \tau_b \circ P \tag{16.13}$$

where $b \in \mathbb{C}^n$ and $P \in GL(\Omega)$. Since Ω is a circular domain, the mapping

$$S_0(z) := -z \tag{16.14}$$

leaves Ω invariant and defines the *symmetry* of Ω about 0. If $\zeta \in \Omega$ is arbitrary, define

$$S_\zeta := \gamma S_0 \gamma^{-1} \tag{16.15}$$

where $\gamma \in \mathrm{Aut}(\Omega)$ satisfies $\gamma(0) = \zeta$. With these symmetry transformations, Ω becomes a hermitean symmetric space of non-compact type [U12, L2].

Now endow $\mathbb{C}^n$ with the unique $GL(\Omega)$-invariant inner product $(z|w)$ which satisfies $(e|e) = r$. Let $dV(z)$ be the corresponding Lebesgue measure. It is known [K3] that the symmetric ball $\Omega \subset \mathbb{C}^n$ (which is different from the "Hilbert" unit ball for the inner product if $r > 1$) has the volume

$$\int_\Omega dV(z) = \pi^n \frac{\Gamma_\Lambda\left(\frac{p}{2}\right)}{\Gamma_\Lambda(p)}, \tag{16.16}$$

where Γ_Λ is the Γ-function introduced in (14.7) and p is the genus. Consider the probability measure

$$d\mu_\Omega(z) = \pi^{-n} \frac{\Gamma_\Lambda(p)}{\Gamma_\Lambda\left(\frac{p}{2}\right)} \cdot dV(z) \tag{16.17}$$

on Ω, and the corresponding L^2-space

$$L^2(\Omega) := L^2(\Omega, d\mu_\Omega) \tag{16.18}$$

of square-integrable functions. $L^2(\Omega)$ is a Hilbert space with inner product

$$(\varphi|\psi)_\Omega := \int_\Omega \overline{\varphi(z)}\psi(z)d\mu_\Omega(z). \tag{16.19}$$

DEFINITION 16.20 The subspace

$$H^2(\Omega) := \{f \in L^2(\Omega) : f \text{ holomorphic}\}$$

of $L^2(\Omega)$ is called the *Bergman space* over Ω.

It is easy to show (cf. [H1]) that $H^2(\Omega)$ is a closed subspace of $L^2(\Omega)$ and is therefore a Hilbert space with inner product (16.19). The orthogonal projection

$$E^\Omega : L^2(\Omega) \to H^2(\Omega) \tag{16.21}$$

is called the *Bergman projection* over Ω. In order to compute the Bergman projection, consider the "Bergman endomorphisms" B(x,y) defined in (16.10). One can show [L2, FK1] that there exists a sesqui-polynomial mapping

$$\Delta : \mathbb{C}^n \times \mathbb{C}^n \to \mathbb{C} \tag{16.22}$$

such that $\Delta(0,0) = 1$ and

$$\text{Det } B(x,y) = \Delta(x,y)^P \tag{16.23}$$

for all $x,y \in Z$. Here p is the genus (14.4). The mapping Δ is called the *Jordan triple determinant.* Note that the triple determinant depends on two arguments, (in a polynomial way on x, in a conjugate polynomial way on y) whereas the Jordan algebra determinant of Z depends only on one variable in Z. It will always be clear from the context whether we use the determinant Δ in the algebra sense or the triple sense. As in (2.48), one can define (16.22) by "Cramer's rule" using the quasi-inverse (16.11) instead of the inverse.

Example 16.24
Let $\mathbb{C}^{r\times r}$ be the space of all complex $(r\times r)$*-matrices, with triple product (16.8). Then it follows from (16.10) that*

$$B(x,y)z = (1 - xy^*)z(1 - y^*x)$$

for all $x,y,z \in Z$, i.e.

$$B(x,y) = L_{1-xy^*} R_{1-y^*x}$$

is a product of left and right multiplication operators. Since

$$\begin{aligned} \text{Det } B(x, y) &= \text{Det}(L_{1-xy^*})\text{Det}(R_{1-y^*x}) \\ &= \text{Det}(1 - xy^*)^r \, \text{Det}(1 - y^*x)^r = \text{Det}(1 - xy^*)^{2r} \end{aligned}$$

and $p = 2n/r = 2r^2/r = 2r$ *we obtain*

$$\Delta(x,y) = \text{Det}(1 - xy^*) \tag{16.25}$$

for all $x,y \in \mathbb{C}^{r\times r}$

PROPOSITION 16.26
The reproducing "Bergman" kernel of $H^2(\Omega)$ *is given by*

$$E^\Omega(z,w) = \Delta(z,w)^{-p} \tag{16.27}$$

for all $z,w \in \Omega$. *Here* Δ *is the Jordan triple determinant and* p *is the genus. Thus for every* $w \in \Omega$, *the holomorphic function*

$$E^\Omega_w(z) := E^\Omega(z,w) \tag{16.28}$$

on Ω *belongs to* $H^2(\Omega)$, *and the Bergman projection (16.21) has the form*

$$(E^\Omega\varphi)(z) = (E^\Omega_z|\varphi)_\Omega \tag{16.29}$$

for all $\varphi \in L^2(\Omega)$ *and* $z \in \Omega$.

PROOF By definition of the Bergman space, we have

$$\text{Det } \partial\gamma(z) \cdot E^\Omega(\gamma(z),\gamma(w)) \cdot \overline{\text{Det } \partial\gamma(w)} = E^\Omega(z,w)$$

for all $z,w \in \Omega$ and all biholomorphic automorphisms γ of Ω. Here $\partial\gamma(z) \in GL(n,\mathbb{C})$ is the complex derivative. Now let $z = \tanh(b) \in \Omega$. The automorphism τ_b defined in (16.12) satisfies $\tau_b(0) = -z$,

$$\partial\tau_b(w) = B(z,z)^{1/2} \, B(w,z)^{-1}$$

and

$$\text{Det } \partial\tau_b(0) = \text{Det } B(z,z)^{1/2} = \Delta(z,z)^{p/2}.$$

Applying (16.30) to $z = w = 0$, we obtain

$$\Delta(z,z)^p \cdot E^\Omega(-z,-z) = E^\Omega(0,0) = 1,$$

where the last equation follows from the normalization. Therefore

$$E^\Omega(z,z) = \Delta(-z,-z)^{-p} = \Delta(z,z)^{-p}.$$

Since (16.27) is sesqui-holomorphic in (z,w), the assertion follows.
Q.E.D.

By (16.30) the measure

$$d\mu(z) := E^\Omega(z,z)\, dV(z) = \Delta(z,z)^{-p}\, dV(z) \tag{16.31}$$

is invariant under the action of $\text{Aut}(\Omega)$. This measure is infinite and does not give rise to an interesting Hilbert space of analytic functions. Let $\lambda > p - 1$ be a parameter and consider the probability measure

$$d\mu_\lambda(z) = \pi^{-n} \frac{\Gamma_\Lambda(\lambda)}{\Gamma_\Lambda\left(\lambda - \frac{p}{2}\right)} \Delta(z, z)^{\lambda - p} dV(z) \tag{16.32}$$

with a normalizing constant involving the Γ-function (14.7). Consider the L^2-space

$$L^2_\lambda(\Omega) := L^2(\Omega, d\mu_\lambda) \tag{16.33}$$

of square-integrable functions on Ω, with inner product

$$(\varphi \mid \psi)_\lambda := \int_\Omega \overline{\varphi(z)}\psi(z) d\mu_\lambda(z). \tag{16.34}$$

For $\lambda = p$, we obtain the space $L^2_p(\Omega) = L^2(\Omega)$ defined in (16.18).

DEFINITION 16.35 The subspace

$$H^2_\lambda(\Omega) := \{f \in L^2_\lambda(\Omega) : f \text{ holomorphic}\}$$

of $L^2_\lambda(\Omega)$ is called the λ-*Bergman space* over Π.

One can show that $H^2_\lambda(\Omega)$ is a closed subspace of $L^2_\lambda(\Omega)$ and is therefore a Hilbert space with inner product $(\varphi|\psi)_\lambda$. The orthogonal projection

$$E^\lambda : L^2_\lambda(\Omega) \to H^2_\lambda(\Omega) \tag{16.36}$$

is called the λ-*Bergman projection* over Ω. In the special case $\lambda = p$, we obtain the standard Bergman space

$$H^2_p(\Omega) = H^2(\Omega)$$

and the standard Bergman projection $E^p = E^\Omega$.

PROPOSITION 16.37
The reproducing "λ-Bergman" kernel of $H^2_\lambda(\Omega)$ *is given by*

$$E^\lambda(z,w) = \Delta(z,w)^{-\lambda} \tag{16.38}$$

for all $z,w \in \Omega$. *Here* Δ *is the Jordan triple determinant and* $\lambda > p - 1$ *is arbitrary.*

Thus for every $w \in \Omega$, the holomorphic function

$$E^\lambda_w(z) := E^\lambda(z,w) \tag{16.39}$$

on Ω belongs to $H^2_\lambda(\Omega)$, and the λ-Bergman projection (16.36) has the form

$$(E^\lambda\varphi)(z) = (E^\lambda_z|\varphi)_\lambda \tag{16.40}$$

for all $\varphi \in L^2_\lambda(\Omega)$ and $z \in \Omega$.

PROPOSITION 16.41
For $\lambda > p - 1$, *there exists a projective unitary representation*

$$U^\Omega_\lambda : G(\Omega) \to U(H^2_\lambda(\Omega)) \qquad (\textit{unitary group})$$

of $G(\Omega)$ *on* $H^2_\lambda(\Omega)$ *satisfying*

$$(U_\lambda^\Omega(\gamma^{-1})f)(z) = f(\gamma(z))(\mathrm{Det}\ \partial\gamma(z))^{\lambda/p} \tag{16.42}$$

for all $\gamma \in G(\Omega)$, $f \in H_\lambda^2(\Omega)$ *and* $z \in \Omega$. *Here, for each* γ, *one chooses a holomorphic branch* $(\mathrm{Det}\ \partial\gamma(z))^{\lambda/p}$ *on the (convex) domain* Ω.

As in the unbounded realization, one may choose the unitary operators $U_\lambda^\Omega(S_\zeta)$, for the symmetries (16.15), to be self-adjoint. With this choice we define

DEFINITION 16.43 Let $f : \Omega \to \mathbb{C}$ be summable for the invariant measure $d\mu$. For each $\lambda > p - 1$, the λ-*Weyl operator* with *active* symbol f is given by the operator integral

$$\sigma_\lambda^*(f) = 2^n \int_\Omega f(\zeta) U_\lambda^\Omega(S_\zeta) d\mu(\zeta) \tag{16.44}$$

on $H_\lambda^2(\Omega)$. Conversely, for any trace-class operator A on $H_\lambda^2(\Omega)$ the smooth function

$$\sigma_\lambda(A)(\zeta) := 2^n\ \mathrm{trace}(A \cdot U_\lambda^\Omega(S_\zeta)) \tag{16.45}$$

on Ω is called the *passive* λ-*Weyl symbol* of A. The two maps σ^*_λ and σ_λ are formally adjoint to each other when viewed as maps between the space of square-summable functions on Ω (with respect to μ) and the space of Hilbert-Schmidt operators on $H_\lambda^2(\Omega)$.

As in Proposition 14.36, one can show that the λ-Weyl calculus in its bounded realization (i.e., realized on Ω) has the covariance group $G(\Omega) = \mathrm{Aut}(\Omega)^0$.

17

Contraction of the λ-Weyl Calculus (Bounded Realization)

The λ-Weyl calculus over the bounded symmetric domain

$$\Omega = \{z \in \mathbb{C}^n : \|z\|_\infty < 1\},$$

realized as the open unit ball of the norm (16.4), admits a contraction as $\lambda \to \infty$ which yields the classical Weyl calculus of pseudodifferential operators (in its complex realization, i.e., realized on the Segal-Bargmann space over $\mathbb{C}^n$). This is parallel to the fact, explained in Section 15, that the λ-Weyl calculus in the unbounded realization Π of Ω contracts to the Fuchs calculus on Λ. In this section we describe the asymptotic behavior, as $\Lambda \to \infty$, of the representation U_λ^Ω of $G(\Omega)$, defined in (16.42), which underlies the λ-Weyl calculus (16.44).

We also need to deform the geometric action of $G(\Omega)$ on the underlying phase space Ω, because the covariance group of the limiting Weyl calculus arises as a contraction of $G(\Omega)$. We will first study the limit $\lambda \to \infty$ on the classical level. Suppose in the following that Ω is a symmetric ball in $\mathbb{C}^n$, with linear automorphism group $GL(\Omega) \subset GL(n,\mathbb{C})$. Consider the biholomorphic mapping $\alpha_\lambda : \Omega \to \sqrt{\lambda}\Omega$ defined by

$$\alpha_\lambda(w) \mapsto \sqrt{\lambda} w. \tag{17.1}$$

Even though nothing distinguishes the geometry of $\sqrt{\lambda}\Omega$ from that of Ω intrinsically, one may think of the map α_λ as flattening things out in the following sense: every compact subset K of $\mathbb{C}^n$ is contained in $\sqrt{\lambda}\Omega$ for large λ; besides, up to a scalar factor depending on λ, the Riemann structure of $\sqrt{\lambda}\Omega$ converges, on K , to that induced by $\mathbb{C}^n$ as $\lambda \to \infty$. Note that, unlike the unbounded version (15.1), α_λ does not preserve Ω as a point set. For every biholomorphic mapping $\gamma \in \mathrm{Aut}(\Omega)$,

$$\mathrm{Ad}(\alpha_\lambda)\gamma := \alpha_\lambda \circ \gamma \circ \alpha_\lambda^{-1} \tag{17.2}$$

defines a biholomorphism of $\sqrt{\lambda}\Omega$.

PROPOSITION 17.3
For the generators of $\mathrm{Aut}(\Omega)$ *and* $z \in \mathbb{C}^n$, *we have, putting,* $b_\lambda := b/\sqrt{\lambda}$,

$$\tilde{P}(z) := \mathrm{Ad}(\alpha_\lambda)(P)(z) = P(z) \qquad (P \in GL(\Omega)), \tag{17.4}$$

$$\tilde{\tau}_b(z) := \lim_{\lambda\to\infty}(\mathrm{Ad}(\alpha_\lambda)\tau_{b_\lambda})(z) = z + b \qquad (b \in \mathbb{C}^n). \tag{17.5}$$

PROOF Since $P \in GL(\Omega)$ is $\mathbb{C}$-linear, we have $\mathrm{Ad}(\alpha_\lambda)P = P$ for every λ. This proves (17.4). Now let $b \in \mathbb{C}^n$ be given and consider the expansion

$$\zeta_\lambda := \tanh\left(\frac{b}{\sqrt{\lambda}}\right) = \sum_{m\geq 0} c_m \frac{b^{2m+1}}{\sqrt{\lambda}^{2m+1}}, \tag{17.6}$$

where c_m are the usual coefficients of tanh about 0, and the odd powers $b^{2m+1} \in \mathbb{C}^n$ are defined in the Jordan triple sense [U12, L2]. Then we have for all $z \in \mathbb{C}^n$ and λ large enough

$$\tau_{b_\lambda}\left(\frac{z}{\sqrt{\lambda}}\right) = \zeta_\lambda + B(\zeta_\lambda, \zeta_\lambda)^{1/2}\left(\frac{z}{\sqrt{\lambda}}\right)^{-\zeta_\lambda}$$

$$= \zeta_\lambda + B(\zeta_\lambda, \zeta_\lambda)^{1/2}B\left(\frac{z}{\sqrt{\lambda}}, -\zeta_\lambda\right)^{-1}\left(\frac{z}{\sqrt{\lambda}} + \left\{\frac{z}{\sqrt{\lambda}}\,\zeta_\lambda^*\,\frac{z}{\sqrt{\lambda}}\right\}\right).$$

This implies

$$\begin{aligned}\mathrm{Ad}(\alpha_\lambda)(\tau_{b_\lambda})(z) &= \sqrt{\lambda}\,\tau_{b_\lambda}\left(\frac{z}{\sqrt{\lambda}}\right)\\ &= \sqrt{\lambda}\,\zeta_\lambda + B(\zeta_\lambda, \zeta_\lambda)^{1/2}B\left(\frac{z}{\sqrt{\lambda}}, -\zeta_\lambda\right)^{-1}\\ &\quad\cdot\left(z + \left\{\frac{z}{\sqrt{\lambda}}\,(\sqrt{\lambda}\,\zeta_\lambda)^*\,\frac{z}{\sqrt{\lambda}}\right\}\right).\end{aligned} \tag{17.7}$$

Since $B(0,0) = \mathrm{id}$, $\zeta_\lambda \to 0$ and $\sqrt{\lambda}\,\zeta_\lambda \to b$, (17.5) follows. Q.E.D.

DEFINITION 17.8 Let G_∞ denote the group of affine transformations of $\mathbb{C}^n$ generated by $GL(\Omega)^\circ \subset GL(n,\mathbb{C})$ and the translation $\tilde{\tau}_b$ $(b \in \mathbb{C}^n)$.

Since $GL(\Omega)$ is a closed (in fact, compact) subgroup of $GL(n,\mathbb{C})$ it follows that G_∞ is a connected Lie group.

PROPOSITION 17.9
G_∞ is a contraction (in the sense of (15.10)) of $G(\Omega)$.

PROOF Since G_∞ is a semi-direct product of the linear group $GL(\Omega)^\circ$ and the translations $\tilde{\tau}_b$ $(b \in \mathbb{C}^n)$, its Lie algebra, denoted by $\mathrm{Lie}(G_\infty)$, is generated by the vector fields

$$\tilde{M} := Mz\,\frac{\partial}{\partial z} \qquad (M \in \mathfrak{gl}(\Omega)), \tag{17.10}$$

$$\tilde{A}_b := b\,\frac{\partial}{\partial z} \qquad (b \in \mathbb{C}^n) \tag{17.11}$$

on $\mathbb{C}^n$, with complex coordinates $z = (z_1, \ldots, z_n)$. Here

$$\mathfrak{gl}(\Omega) := \{M \in \mathfrak{gl}(n, \mathbb{C}) : \exp(\theta M)\Omega = \Omega\ \forall \theta \in \mathbb{R}\}$$

is the Lie algebra of $GL(\Omega)$, and $\partial/\partial z$ is a column vector. The commutator (15.22) has the form

$$[\tilde{M}, \tilde{A}_b] = \tilde{A}_{Mb} \qquad (b \in \mathbb{C}^n, M \in \mathfrak{gl}(\Omega)).$$

and

$$[\tilde{M}_1,\tilde{M}_2] = (M_1M_2 - M_2M_1)^\sim \qquad (M_1,M_2 \in \mathfrak{gl}(\Omega)).$$

Trivially, we have for all $b,c \in \mathbb{C}^n$

$$[\tilde{A}_b,\tilde{A}_c] = 0.$$

This describes the Lie algebra $\mathrm{Lie}(G_\infty)$.

Now consider the complex domain Ω. By Cartan's Theorem, the Lie algebra $\mathfrak{g} := \mathrm{aut}(\Omega)$ of $G(\Omega)$ consists of all completely integrable holomorphic vector fields

$$A = h(z)\,\frac{\partial}{\partial z}$$

on Ω. Here $h : \Omega \to \mathbb{C}^n$ is a holomorphic mapping, and complete integrability means that there exists a continuous 1-parameter group $\gamma_\theta = \exp(\theta A) \in G(\Omega)$ satisfying the ordinary differential equation

$$\frac{\partial \gamma_\theta(z)}{\partial \theta} = h(\gamma_\theta(z))$$

for all $\theta \in \mathbb{R}$ and $z \in \Omega$. The holomorphic vector fields

$$\hat{M} = Mz\,\frac{\partial}{\partial z} \qquad (M \in \mathfrak{gl}(\Omega)) \tag{17.12}$$

and

$$A_b = (b - \{zb^*z\})\,\frac{\partial}{\partial z} \qquad (b \in \mathbb{C}^n) \tag{17.13}$$

on Ω satisfy

$$\exp(\hat{M}) = \exp(M)$$

and

$$\exp(A_b) = \tau_b.$$

It follows that these vector fields are completely integrable on Ω. We put

$$\mathfrak{k} := \{\hat{M} : M \in \mathfrak{gl}(\Omega)\}$$

and

$$p := \{A_b : b \in \mathbb{C}^n\}.$$

Then it is known [U12, L2] that

$$\mathfrak{g} = \mathrm{aut}(\Omega) = \mathfrak{k} \oplus p \tag{17.14}$$

Clearly, $\mathfrak{k}$ is a subalgebra of $\mathfrak{g}$ (consisting of all completely integrable *linear* vector fields on Ω), whereas p is only a subspace of $\mathfrak{g}$. Let $\mathfrak{g}_c$ be the contracted Lie algebra structure on $\mathfrak{g}$ associated with the splitting (17.14). For the commutator (15.22) of vector fields on Ω we get

$$[\hat{M}, A_b] = (M(b - \{zb^*z\}) + 2\{(Mz)b^*z\}) \frac{\partial}{\partial z}$$

$$= A_{Mb} \qquad (b \in \mathbb{C}^n, M \in \mathfrak{gl}(\Omega)),$$

since $M\{zb^*z\} = 2\{(Mz)b^*z\} + \{z(Mb)^*z\}$, and

$$[\hat{M}_1,\hat{M}_2] = (M_1M_2 - M_2M_1)^\wedge \qquad (M_1,M_2 \in \mathfrak{gl}(\Omega)).$$

Since for all $b,c \in \mathbb{C}^n$

$$[A_b, A_c] \in [p,p] \subset \mathfrak{k}.$$

it follows that $\mathrm{Lie}(G_\infty)$ is isomorphic to the contraction $\mathfrak{g}_c$ of $\mathfrak{g} = \mathrm{aut}(\Omega)$ by sending $\tilde{A}_b \mapsto A_b$ and $\tilde{M} \mapsto \hat{M}$. Q.E.D.

We now pass to the study of the representation U_λ^Ω of $G = \mathrm{Aut}(\Omega)^\circ$, as $\lambda \to \infty$. The first step is a "flat" realization of the weighted Bergman spaces $H_\lambda^2(\Omega)$. This involves some harmonic analysis. Consider the algebra

$$\mathcal{P}(\mathbb{C}^n) = \sum_{m \in N}^{\oplus} \mathcal{P}^m(\mathbb{C}^n) \tag{17.15}$$

of all (holomorphic) polynomials $\varphi : \mathbb{C}^n \to \mathbb{C}$, endowed with its natural gradation into the finite-dimensional subspaces $\mathcal{P}^m(\mathbb{C}^n)$ consisting of all m-homogeneous polynomials. Let $(z|w)$ be the $GL(\Omega)$-invariant inner product on $\mathbb{C}^n$ normalized by $(e|e) = r$ and consider the *Segal-Bargmann space* $H^2(\mathbb{C}^n)$ of all holomorphic functions $\varphi : \mathbb{C}^n \to \mathbb{C}$ such that

$$\|\varphi\|^2 := \frac{1}{\pi^n} \int_{\mathbb{C}^n} e^{-(z|z)} |\varphi(z)|^2 dV(z) < \infty. \tag{17.16}$$

Let $(\varphi|\psi)$ denote the inner product in $H^2(\mathbb{C}^n)$. By Taylor's expansion, $\mathcal{P}(\mathbb{C}^n)$ is a dense subspace of $H^2(\mathbb{C}^n)$. It is easy to show that

$$(\varphi|\psi) = (\partial_\varphi \psi)(0) \tag{17.17}$$

for all $\varphi, \psi \in \mathcal{P}(\mathbb{C}^n)$. Here ∂_φ denotes the constant coefficient differential operator on $\mathbb{C}^n$ canonically associated with φ via the hermitian structure

(∂_φ depends in a conjugate-linear way on φ). The group $GL(\Omega)$ acts on $H^2(\mathbb{C}^n)$ by the unitary transformations

$$\rho_P(\varphi)(z) = \varphi(P^{-1}z) \qquad (P \in GL(\Omega)) \tag{17.18}$$

for all $\varphi \in H^2(\mathbb{C}^n)$ and $z \in \mathbb{C}^n$. This action leaves $\mathcal{P}(\mathbb{C}^n)$ invariant. We will also consider the complexification $GL(\Omega)^{\mathbb{C}}$ and its complexified action (17.18). A basic problem is to decompose $H^2(\mathbb{C}^n)$ [or its dense subspace $\mathcal{P}(\mathbb{C}^n)$] into *irreducible* invariant subspaces.

DEFINITION 17.19 Let $\mathbb{N}^r_+$ be the set of all sequences

$$\underline{m} = (m_1, \ldots, m_r) \tag{17.20}$$

of integers $m_1 \geq m_2 \geq \cdots \geq m_r \geq 0$. The elements of $\mathbb{N}^r_+$ are called *partitions* or *signatures.*

Let $e_1, \ldots, e_r$ be a *frame* of the irreducible Jordan algebra $\mathbb{R}^n$. Let

$$\mathbb{R}^n = \sum_{1\leq i\leq j\leq r} X_{ij} \tag{17.21}$$

be the associated Peirce decomposition (cf. [L2, FK2]). We have dim X_{ii} = 1 and dim X_{ij} = (i < j), accounting for the formula (3.3). For each $1 \leq \ell \leq r$,

$$X_\ell := \sum_{1\leq i\leq j\leq \ell} X_{ij} \tag{17.22}$$

is a subalgebra of $\mathbb{R}^n$ with unit element $e_1 + \cdots + e_\ell$. Let $P_\ell : \mathbb{R}^n \to X_\ell$ be the orthogonal projection.

DEFINITION 17.23 The ℓth *minor* is the polynomial

$$\Delta_\ell : \mathbb{R}^n \to \mathbb{R} \tag{17.23}$$

obtained by

$$\Delta_\ell(x) := \Delta_{X_\ell}(P_\ell x) \tag{17.24}$$

for all x, where Δ_{X_ℓ} is the Jordan algebra determinant of (17.22). By complexification, Δ_ℓ becomes a polynomial of degree ℓ on $\mathbb{C}^n$. For example

$$\Delta_r(z) = \Delta(z)$$

is the Jordan algebra determinant, whereas $\Delta_1(z) = (z|e_1)$ for all $z \in \mathbb{C}^n$. Note that $\Delta_1, \ldots, \Delta_{r-1}$ depend on the ordering of the frame.

THEOREM 17.25
For each signature $\underline{m} \in \mathbb{N}^r_+$*, define the "conical polynomial"*

$$\Delta_{\underline{m}}(z) := \Delta_1(z)^{m_1-m_2}\Delta_2(z)^{m_2-m_3} \cdots \Delta_r(z)^{m_r}. \tag{17.26}$$

Let $\mathcal{P}^{\underline{m}}(\mathbb{C}^n)$ *be the* $GL(\Omega)$*-submodule of* $\mathcal{P}(\mathbb{C}^n)$ *generated by* $\Delta_{\underline{m}}$*, i.e., the linear span of all translations* $\rho_P(\Delta_{\underline{m}})$*, where* $P \in GL(\Omega)$*. Then each* $\mathcal{P}_{\underline{m}}(\mathbb{C}^n)$ *is irreducible with highest weight vector* $\Delta_{\underline{m}}$*, and there is an orthogonal decomposition*

$$\mathcal{P}(\mathbb{C}^n) = \sum_{\underline{m} \in \mathbb{N}^r_+} \mathcal{P}^{\underline{m}}(\mathbb{C}^n) \tag{17.27}$$

into pairwise inequivalent modules.

PROOF On a representation theoretic level, this theorem was proved in [S1] by identifying the highest weights in $\mathcal{P}(\mathbb{C}^n)$ and their multiplicity. The explicit description (17.26) of the highest weight vectors was done in [U14] (cf. also [J1]). Q.E.D.

For each $\underline{m} \in \mathbb{N}^r_+$, the polynomials $\psi \in \mathcal{P}^{\underline{m}}(\mathbb{C}^n)$ will be called (pure) of type $\underline{m}$. These polynomials are homogeneous of degree

$$|\underline{m}| := m_1 + \cdots + m_r.$$

Hence the type is a combinatorial refinement ("symmetric degree") of the ordinary degree of polynomials. For each $\underline{m} \in \mathbb{N}^r_+$, let $E^{\underline{m}}(z,w)$ be the reproducing kernel of the subspace $\mathcal{P}^{\underline{m}}(\mathbb{C}^n)$ of $H^2(\mathbb{C}^n)$. This means that for each $w \in \mathbb{C}^n$, the function

$$E^{\underline{m}}_z(w) := E^{\underline{m}}(z,w) \tag{17.28}$$

is a polynomial of type $\underline{m}$, and we have

$$\varphi(z) = (E^{\underline{m}}_z|\varphi) \tag{17.29}$$

for every $\varphi \in \mathcal{P}^{\underline{m}}(\mathbb{C}^n)$ and $z \in \mathbb{C}^n$. Taking the completion of the decomposition (17.27), we obtain

PROPOSITION 17.30
The Segal-Bargmann reproducing kernel

$$E(z,w) = e^{(z|w)} \tag{17.31}$$

has a compactly convergent series expansion

$$E(z,w) = \sum_{\underline{m} \in N_+^r} E^{\underline{m}}(z,w). \tag{17.32}$$

REMARK 17.33
The kernel functions (17.28) are related to the so-called "spherical" polynomials of type $\underline{m}$ [cf. (17.39) below]. However, these polynomials have a more complicated description than the conical polynomials (17.26).

DEFINITION 17.34 Define the *multi-Pochhammer symbol*

$$((\lambda))_{\underline{m}} = \prod_{j=1}^{r} \left(\lambda - \frac{a}{2}(j-1)\right)_{m_j} \tag{17.35}$$

as a polynomial of degree $|\underline{m}|$ in $\lambda \in \mathbb{C}$. Here a is the characteristic multiplicity of the Jordan algebra $\mathbb{R}^n$, and

$$(\alpha)_m := \alpha(\alpha + 1) \cdots (\alpha + m - 1) \tag{17.17}$$

is the usual Pochhammer symbol.

THEOREM 17.36
For each $\lambda > p - 1$, *there exists a Hilbert space isomorphism*

$$\mathcal{F}_\lambda: H_\lambda^2(\Omega) \underset{\cong}{\to} H^2(\mathbb{C}^n) \tag{17.18}$$

with adjoint

$$\mathcal{L}_\lambda(\varphi_{\underline{m}}) = \sqrt{((\lambda))_{\underline{m}}} \cdot \varphi_{\underline{m}} \tag{17.37}$$

for every signature $\underline{m} \in N_+^r$ and every polynomial $\varphi_{\underline{m}} \in \mathcal{P}^{\underline{m}}(\mathbb{C}^n)$ of type $\underline{m}$. *Here* $((\lambda))_{\underline{m}}$ *is the multi-Pochhammer symbol.*

PROOF Let $e := e_1 + \cdots e_r$ and consider the subgroup

$$L := \{P \in GL(\Omega) : Pe = e\} \tag{17.38}$$

of $GL(\Omega)$, endowed with the normalized Haar measure $d\ell$. The polynomial

$$\phi_{\underline{m}}(z) := \int_L \Delta_{\underline{m}}(\ell \cdot z)\, d\ell \tag{17.39}$$

is called the spherical polynomial of type $\underline{m}$, it is the unique L-invariant polynomial of type $\underline{m}$ normalized by $\phi_{\underline{m}}(e) = 1$. Since the reproducing kernel vector $E_e^{\underline{m}}$ of type $\underline{m}$ is also L-invariant, it is a multiple of $\phi_{\underline{m}}$. If x belongs to the strictly positive cone Λ of $\mathbb{R}^n$, we have

$$\begin{aligned} E^{\underline{m}}(x^2,e) &= (E_{x^2}^{\underline{m}} \mid E_e^{\underline{m}}) = (\rho_P E_x^{\underline{m}} \mid E_e^{\underline{m}}) \\ &= (E_x^{\underline{m}} \mid \rho_P E_e^{\underline{m}}) = (E_x^{\underline{m}} \mid E_x^{\underline{m}}) = E^{\underline{m}}(x,x). \end{aligned}$$

Here $P = P_x^{1/2}$ is the quadratic representation operator which belongs to the complexification of $GL(\Omega)$. By (17.17), ρ_P is self adjoint on $H^2(\mathbb{C}^n)$. For the normalized Haar measure dP on $GL(\Omega)$ it follows that

$$\begin{aligned} &\int_{GL(\Omega)} |E_e^{\underline{m}}(Px)|^2 dP \\ &= \int_{GL(\Omega)} |(E_{Px}^{\underline{m}} \mid E_e^{\underline{m}})|^2 dP = \int_{GL(\Omega)} |(\rho_{P^{-1}} E_x^{\underline{m}} \mid E_e^{\underline{m}})|^2 dP \\ &= \frac{1}{d_{\underline{m}}} (E_x^{\underline{m}} \mid E_x^{\underline{m}})(E_e^{\underline{m}} \mid E_e^{\underline{m}}) = \frac{1}{d_{\underline{m}}} E^{\underline{m}}(x^2,e) E^{\underline{m}}(e,e). \end{aligned}$$

Here

$$d_{\underline{m}} := \dim \mathcal{P}^{\underline{m}}(\mathbb{C}^n)$$

and we applied Schur orthogonality to the irreducible modules $\mathcal{P}^{\underline{m}}(\mathbb{C}^n)$. By proportionality, we obtain

$$\int_{GL(\Omega)} |\phi_{\underline{m}}(Px)|^2 dP = \frac{1}{d_{\underline{m}}} \phi_{\underline{m}}(x^2) \tag{17.40}$$

for all $x \in \Lambda$. Now we compute the norm of $\phi_{\underline{m}}$ in $H^2(\mathbb{C}^n)$. Let $\equiv$ denote equality up to a factor independent of $\underline{m}$. Integrating in polar coordinates

[FK1, FK2] and changing variables $y_j = x_j^2$, we obtain from (17.16) and (17.40)

$$\begin{aligned} d_{\underline{m}}(\phi_{\underline{m}} | \phi_{\underline{m}}) &\equiv d_{\underline{m}} \int_{C^n} e^{-(z|z)} |\phi_{\underline{m}}(z)|^2 \, dV(z) \\ &\equiv d_{\underline{m}} \int_{R_+^r} \int_{GL(\Omega)} \left| \phi_{\underline{m}}\left(P \cdot \sum_{j=1}^r x_j e_j \right) \right|^2 \cdot e^{-\sum_{j=1}^r x_j^2} \\ &\quad \cdot \prod_{1 \le i < j \le r} |x_i^2 - x_j^2|^a \cdot \prod_{j=1}^r x_j \, dP \, dx_1 \cdots dx_r \\ &= \int_{R_+^r} \phi_{\underline{m}}\left(\sum_{j=1}^r x_j^2 e_j \right) \cdot e^{-\sum_{j=1}^r x_j^2} \cdot \prod_{1 \le i < j \le r} |x_i^2 - x_j^2|^a \prod_{j=1}^r x_j \, dx_j \cdots dx_r \\ &\equiv \int_{R_+^r} \phi_{\underline{m}}\left(\sum_{j=1}^r y_j e_j \right) \cdot e^{-\sum_{j=1}^r y_j} \cdot \prod_{1 \le i < j \le r} |y_i - y_j|^a \, dy_1 \cdots dy_r \\ &= \int_\Lambda \phi_{\underline{m}}(y) e^{-(y|e)} \, dy = \int_\Lambda \Delta_{\underline{m}}(y) e^{-(y|e)} \, dy \equiv \Gamma_\Lambda\left(\underline{m} + \frac{n}{r} \right) \end{aligned}$$

by a result of Gindikin (cf. [FK2, G1]). In order to compute the $H_\lambda^2(\Omega)$-norm of a $\phi_{\underline{m}}$, we obtain, using [FK1]

$$\begin{aligned} d_{\underline{m}}(\phi_{\underline{m}} | \phi_{\underline{m}})_\lambda &\equiv d_{\underline{m}} \int_\Omega |\phi_{\underline{m}}(z)|^2 \Delta(z,z)^{\lambda - p} \, dV(z) \\ &\equiv d_{\underline{m}} \int_{[0,1)^r} \int_{GL(\Omega)} \left| \phi_{\underline{m}}\left(P \cdot \sum_{j=1}^r x_j e_j \right) \right|^2 \\ &\quad \cdot \Delta\left(P \cdot \sum_{j=1}^r x_j e_j, P \cdot \sum_{j=1}^r x_j e_j \right)^{\lambda - p} \\ &\quad \cdot \prod_{1 \le i < j \le r} |x_i^2 - x_j^2|^a \prod_{j=1}^r x_j \, dP \, dx_1 \cdots dx_r \\ &\equiv \int_{[0,1)^r} \phi_{\underline{m}}\left(\sum_{j=1}^r x_j^2 e_j \right) \prod_{j=1}^r (1 - x_j^2)^{\lambda - p} \\ &\quad \cdot \prod_{1 \le i < j \le r} |x_i^2 - x_j^2|^a \prod_{j=1}^r x_j \, dx_1 \cdots dx_r \\ &\equiv \int_{[0,1)^r} \phi_{\underline{m}}\left(\sum_{j=1}^r y_j e_j \right) \prod_{j=1}^r (1 - y_j)^{\lambda - p} \end{aligned}$$

$$\cdot \prod_{1\le i<j\le r} |y_i - y_j|^a \, dy_1 \cdots dy_r$$

$$\equiv \int_{\Lambda\cap(e-\Lambda)} \phi_{\underline{m}}(y)\Delta(e - y)^{\lambda-p}\, dy$$

$$= \int_{\Lambda\cap(e-\Lambda)} \Delta_{\underline{m}}(y)\Delta(e - y)^{\lambda-p}\, dy$$

$$\equiv \frac{\Gamma_\Lambda\left(\underline{m} + \frac{n}{r}\right)}{\Gamma_\Lambda(\underline{m} + \lambda)}$$

by a result of Gidikin [FK2, G1]. Comparing these two expressions and evaluating for $\underline{m} = (0, \ldots, 0)$, we see that

$$\frac{(\phi_{\underline{m}} | \phi_{\underline{m}})}{(\phi_{\underline{m}} | \phi_{\underline{m}})_\lambda} = \frac{\Gamma_\Lambda(\underline{m} + \lambda)}{\Gamma_\Lambda(\lambda)} = ((\lambda))_{\underline{m}}.$$

Since the invariant inner products $(|)$ and $(|)_\lambda$ on $\mathcal{P}^{\underline{m}}(\mathbb{C}^n)$ are proportional, we obtain

$$(\varphi|\psi) = ((\lambda))_{\underline{m}}\, (\varphi|\psi)_\lambda \tag{17.41}$$

for all $\varphi, \psi \in \mathcal{P}^{\underline{m}}(\mathbb{C}^n)$. Since the polynomials are dense in $H^2(\mathbb{C}^n)$, and pure polynomials of different type are orthogonal with respect to both inner products, it follows that $\mathcal{L}_\lambda$ is an isometry. Since the domain Ω is invariant under the circle action, every holomorphic function on Ω has an expansion into a series of m-homogeneous polynomials which converges compactly on Ω. Therefore $\mathcal{P}(\mathbb{C}^n)$ is dense in $H^2_\lambda(\Omega)$ showing that $\mathcal{L}_\lambda$ is surjective.

Q.E.D.

DEFINITION 17.42 For $\lambda > p - 1$, define a projective representation

$$U_\lambda : G(\Omega) \to U(H^2(\mathbb{C}^n)) \qquad \text{(unitary group)}$$

of $G(\Omega)$ on the Segal-Bargmann space $H^2(\mathbb{C}^n)$ by putting

$$U_\lambda(\gamma) := \mathcal{L}_\lambda^* \, U_\lambda^\Omega(\gamma)\, \mathcal{L}_\lambda \tag{17.43}$$

for all $\gamma \in G(\Pi)$. Here $\mathcal{L}_\lambda$ is defined by (17.37).

PROPOSITION 17.44
The group G_∞ *defined in* (17.8) *has a unitary representation*

$$U_\infty : G_\infty \to U(H^2(\mathbb{C}^n)) \qquad \textit{(unitary group)}$$

on $H^2(\mathbb{C}^n)$ *given by*

$$(U_\infty(\tilde{\tau}_b)\varphi)(z) = \varphi(z - b)\, e^{(z|b)-(b|b)/2} \qquad (b \in \mathbb{C}^n), \tag{17.45}$$

$$(U_\infty(\tilde{P})\varphi)(z) = \varphi(P^{-1}z) \qquad (P \in GL(\Omega)). \tag{17.46}$$

PROOF G_∞ is a subgroup of the semi-direct product $U(n) \times \mathbb{C}^n$ which has a representation of this form. Q.E.D.

THEOREM 17.47
For the generators of G_∞, *we have*

$$U_\infty(\tilde{P}) = U_\lambda(P) \qquad (P \in GL(\Omega)), \tag{17.48}$$

$$U_\infty(\tilde{\tau}_b) = \lim_{\lambda\to\infty} U_\lambda(\tau_{b_\lambda}) \qquad (b \in \mathbb{C}^n), \tag{17.49}$$

where $b_\lambda := b/\sqrt{\lambda}$.

PROOF If $P \in GL(\Omega)$ and $\varphi \in \mathcal{P}(Z)$ is a polynomial, we have for all $z \in \mathbb{C}^n$

$$(U_\lambda(P)\varphi)(z) = \varphi(P^{-1}z) = U_\infty(\tilde{P})\varphi(z)$$

independently of λ. This follows from the fact that $\mathcal{L}_\lambda$ defined in (17.37) intertwines the canonical actions of $GL(\Omega)$ on $H^2_\lambda(\Omega)$ and $H^2(\mathbb{C}^n)$, respectively. In order to study the limit (17.49), we pass to the Lie algebra level, and consider the infinitesimal generator

$$dU^\Omega_\lambda : \mathfrak{g} \to \mathcal{U}(H^2_\lambda(\Omega)) \tag{17.50}$$

of U^Ω_λ. Here the Lie algebra $\mathfrak{g} = \mathrm{aut}(\Omega)$ acts by essentially skew-adjoint operators defined on a dense subspace of $H^2_\lambda(\Omega)$ (including the polynomials). By differentiating (16.42) we see that for every vector field $A = h(z)\,\partial/\partial z \in \mathfrak{g}$ we have

$$dU_\lambda^\Omega(A)f(z) = \partial f(z)h(z) - \frac{\lambda}{p} f(z) \text{ trace } \partial h(z).$$

If $h(z) = b - \{zb^*z\}$, then

$$\partial h(z)w = -2\{zb^*w\}$$

for all $w \in \mathbb{C}^n$ and therefore

$$\text{trace } \partial h(z) = -p \cdot (z|b).$$

It follows that

$$dU_\lambda^\Omega\left((b - \{zb^*z\}) \frac{\partial}{\partial z}\right)f(z) = -\partial f(z)b + \partial f(z)\{zb^*z\} + \lambda f(z)(z|b).$$

For any signature $\underline{m} \in \mathbb{N}_+^r$ and any polynomial $\varphi \in \mathcal{P}(\mathbb{C}^n)$ let

$$\varphi_{\underline{m}} \in \mathcal{P}^{\underline{m}}(\mathbb{C}^n)$$

denote the "component" of φ of type $\underline{m}$. Also put

$$\underline{m} \pm \varepsilon_j := (m_1, \ldots, m_{j-1}, m_j \pm 1, m_{j+1}, \ldots, m_r) \tag{17.52}$$

if $1 \leq j \leq r$, provided these multi-indices give rise to (non-increasing) partitions in $\mathbb{N}_+^r$. With this notation, we can determine the infinitesimal generator

$$dU_\lambda : \mathfrak{g} \to \mathcal{U}(H^2(\mathbb{C}^n)) \tag{17.53}$$

of U_λ as follows: Let $\varphi \in \mathcal{P}^{\underline{m}}(\mathbb{C}^n)$ for some signature $\underline{m} \in \mathbb{N}_+^r$. By [U15; Corollary 2.10 and Theorem 2.11], we have

$$\partial\varphi(z)b = \sum_{j=1}^{r} (\partial\varphi(z)b)_{\underline{m}-\epsilon_j},$$

$$\partial\varphi(z)\{zb^*z\} = \sum_{j=1}^{r} (\partial\varphi(z)\{zb^*z\})_{\underline{m}+\epsilon_j},$$

$$\varphi(z)\cdot(z|b) = \sum_{j=1}^{r} (\varphi(z)\cdot(z|b))_{\underline{m}+\epsilon_j},$$

$$(\partial\varphi(z)\{zb^*z\})_{\underline{m}+\epsilon_j} = \left(m_j - \frac{a}{2}(j-1)\right)\cdot(\varphi(z)\cdot(z\,|\,b))_{\underline{m}+\epsilon_j}$$

for every j. Therefore

$$\begin{aligned}
&dU_\lambda\left((b - \{zb^*z\})\frac{\partial}{\partial z}\right)\varphi(z)\\
&= \sqrt{((\lambda))_{\underline{m}}}\,\mathcal{F}_\lambda(\partial\varphi(z)\{zb^*z\} + \lambda\varphi(z)\cdot(z\,|\,b) - \partial\varphi(z)b)\\
&= \sqrt{((\lambda))_{\underline{m}}}\\
&\quad\cdot\sum_{j=1}^{r}\left\{\frac{1}{\sqrt{((\lambda))_{\underline{m}+\epsilon_j}}}\left(\lambda + m_j - \frac{a}{2}(j-1)\right)(\varphi(z)\cdot(z\,|\,b))_{\underline{m}+\epsilon_j}\right.\\
&\quad\left. - \frac{1}{\sqrt{((\lambda))_{\underline{m}-\epsilon_j}}}(\partial\varphi(z)b)_{\underline{m}-\epsilon_j}\right\}\\
&= \sum_{j=1}^{r}\left\{\sqrt{\lambda + m_j - \frac{a}{2}(j-1)}(\varphi(z)\cdot(z\,|\,b))_{\underline{m}+\epsilon_j}\right.\\
&\quad\left. - \sqrt{\lambda + m_j - 1 - \frac{a}{2}(j-1)}(\partial\varphi(z)b)_{\underline{m}-\epsilon_j}\right\}.
\end{aligned}$$

Replacing b by $b/\sqrt{\lambda}$, we obtain for $\varphi \in \mathcal{P}^{\underline{m}}(\mathbb{C}^n)$

$$\begin{aligned}
&dU_\lambda\left(\left(\frac{b}{\sqrt{\lambda}} - \left\{z\left(\frac{b}{\sqrt{\lambda}}\right)^* z\right\}\right)\frac{\partial}{\partial z}\right)\varphi(z)\\
&\quad= \sum_{j=1}^{r}\left\{\left(\frac{\lambda + m_j - \frac{a}{2}(j-1)}{\lambda}\right)^{1/2}\cdot(\varphi(z)(z|b))_{\underline{m}+\varepsilon j}\right.\\
&\quad\left. - \left(\frac{\lambda + m_j - 1 - \frac{a}{2}(j-1)}{\lambda}\right)^{1/2}(\partial\varphi(z)b)_{m-\varepsilon j}\right\}
\end{aligned}$$

As $\lambda \to \infty$, we get

$$\lim_{\lambda\to\infty} dU_\lambda\left(\left(\frac{b}{\sqrt{\lambda}} - \left\{z\left(\frac{b}{\sqrt{\lambda}}\right)^* z\right\}\right)\frac{\partial}{\partial z}\,\varphi(z)\right.$$

$$= \sum_{j=1}^{r} ((\varphi(z)\cdot(z|b))_{\underline{m}+\epsilon_j} - (\partial\varphi(z)b)_{\underline{m}-\epsilon_j}) = \varphi(z)\cdot(z|b) - \partial\varphi(z)b. \quad (17.54)$$

This limit is independent of the type $\underline{m}$ and thus holds for all polynomials $\varphi \in \mathcal{P}(\mathbb{C}^n)$. Since (17.54) is the infinitesimal generator of the translation action the assertion follows. Q.E.D.

References

[BG] Baouendi, M.S. and Goulaouic, C., Régularité et théorie spectrale pour une classe d'opérateurs elliptiques dégénérés, *Arch. Rat. Mech. Anal.* 34 (1970), 361–379.

[B1] Beals, R., A general calculus of pseudodifferential operators, *Duke Math. J.* 42 (1975), 1–42.

[B2] Beals, R., Characterization of pseudodifferential operators and applications, *Duke Math. J.* 44 (1977), 45–57.

[BF] Beals, R. and Fefferman, C., On local solvability of linear partial differential equations. *Ann. Math.* 97 (1973), 482–498.

[B3] Berezin, F.A., Quantization, *Math. USSR Izvest.* 8 (1974), 1109–1165.

[B4] Berezin, F.A., Quantization in complex symmetric spaces, *Math. USSR Izvest.* 9 (1975), 341–379.

[BC] Bolley, P., and Camus, J., Sur une classe d'opérateurs elliptiques et dégénérés à plusieurs variables, *Bull. Soc. Math. France,* Mémoires #34 (1973), 55–140.

[BCD] Bolley, P., Camus, and J., Dauge, M., Régularité Gevrey pour le problème de Dirichlet dans des domaines à singularités coniques. *Comm. Part. Diff. Equ.* 10 (1985), 391–431.

[B5] Bourdaud, G., Une algébre maximale d'opérateurs pseudo-différentiels, *Comm. Part. Diff. Equ.* 13 (1988), 1059–1084.

[B6] Boutet de Monvel, L., Boundary problems for pseudo-differential operators, *Acta Math.* 126 (1971), 11–51.

[B7] Braun, H., Hermitian modular functions, *Ann. Math.* 50 (1949), 827–855.

[BK] Braun, H. and Koecher, M., *Jordan-Algebren,* Springer-Verlag, Berlin, 1966.

[B8] Bruyant, F., Estimations pour la composition d'un grand nombre d'opérateurs pseudo-différentiels et applications, Thèse, Univ. de Reims (1979).

[C1] Calderon, A.P., Uniqueness in the Cauchy problem of partial differential equations, *Amer. J. Math.* 80 (1958), 16–36.

[C2] Calderon, A.P., Intermediate spaces and interpolation, the complex method, *Studia Math.* 24 (1964), 113–190.

[CV] Calderon, A.P. and Vaillancourt, R., On the boundedness of pseudodifferential operators, *J. Math. Soc. Japan* 23 (1971), 374–378.

[CZ] Calderon, A.P. and Zygmund, A., Singular integral operators and differential equations, *Am. J. Math.* 79 (1957), 901–921.

[C3] Cartier, P., Quantum mechanical commutation relations and theta functions, *Proc. Symp. Pure Math.* 9 (1966), 361–383.

[CF] Cordoba, A. and Fefferman, C., Wave packets and Fourier integral operators, *Comm. Part. Diff. Equ.* 3 (1978), 979–1006.

[D1] Dauge, M., *Elliptic Boundary Value Problems on Corner Domains, Lect. Notes in Math. 1341,* Springer-Verlag, Berlin, 1988.

[D2] De Bruijn, N.G., Uncertainty principles in Fourier analysis, *Proc. Symp. on Inequalities,* Academic Press, New York, 1967, 57–71.

[DK] Dorfmeister, J. and Koecher, M., Reguläre Kegel, *Jber. Dt. Math. Verein.* 81 (1979), 109–151.

[E1] Egorov, Y., On canonical transformations of pseudodifferential operators, *Uspek Mat. Nauk.* 25 (1969), 235–236.

[F1] Faraut, J., *Algèbres de Jordan et cônes symétriques, Prépublications,* Univ. de Poitiers, Poitiers, France, 1988.

[FG] Faraut, J. and Gindikin, S., Deux formules d'inversion pour la transformation de Laplace sur un cône symétrique, *C. R. Acad. Sci.* Paris 310 (1990), 5–8.

[FK1] Faraut, J. and Korányi, A., Function spaces and reproducing kernels on bounded symmetric domains, *J. Funct. Anal.* 88 (1990), 64–89.

[FK2] Faraut, J. and Korányi, A., *Analysis on Symmetric Cones,* Oxford Univ. Press, Oxford, 1994.

[F2] Folland, G.B., *Harmonic Analysis in Phase Space, Ann. Math. Studies 122,* Princeton Univ. Press, Princeton, 1990.

[G1] Gindikin, S., Analysis on homogeneous domains. *Russ. Math. Surv.* 19 (1964), 3–22.

[G2] Grubb, G., *Functional Calculus of Pseudo-Differential Boundary Problems, Progress in Math. 65,* Birkhäuser, Boston, 1986.

[GK] Gross, K.I. and Kunze, R.A., Fourier Bessel transforms and holomorphic discrete series, *Lect. Notes in Math.* 266 (1972), 79–122.

[H1] Helgason, S., *Differential Geometry and Symmetric Spaces,* Academic Press, New York, 1962.

[H2] Hörmander, L., Pseudo-differential operators and hypoelliptic equations, *Proc. Symp. Pure Math.* 10 (1966), 138–183.

[H3] Hörmander, L., On the L^2-continuity of pseudodifferential operators, *Comm. Pure Appl. Math.* 24 (1971), 529–535.

[H4] Hörmander, L., The Weyl calculus of pseudodifferential operators, *Comm. Pure Appl. Math.* 32 (1979), 359–443.

[H5] Hörmander, L., The Analysis of Linear Partial Differential Operators III, Springer-Verlag, Berlin, 1985.

[H6] Howe, R., The oscillator semi-group, *Proc. Symp. Pure Math.* 48 (1988), 61–132.

[H7] Hirzebruch, U. and Halbräume und ihre holomorphen Automorphismen, *Math. Ann.* 153 (1964), 395–417.

[I1] Igusa, J., Theta Functions, Springer-Verlag, Berlin, 1972.

[J1] Johnson, K.D., On a ring of invariant polynomials on a hermitian symmetric space, *J. Alg.* 67 (1980), 72–81.

[JNW] Jordan, P., von Neumann, J., and Wigner, E., On an algebraic generalization of the quantum mechanical formalism, *Ann. Math.* 36 (1934), 29–64.

[KN] Kohn, J. and Nirenberg, L., An algebra of pseudo-differential operators, *Comm. Pure Appl. Math.* 18 (1965), 269–305.

[K1] Kondrat'ev, V.A., Boundary value problems in domains with conical or angular points, *Trans. Moscow Math. Soc.* 16 (1967), 227–313.

[K2] Korányi, A., *Complex Analysis and Symmetric Domains, Prépublications,* Univ. de Poitiers, France, 1988.

[K3] Korányi, A., The volume of symmetric domains, the Koecher Gamma function and an integral of Selberg, *Studia Scient. Math. Hung.* 17 (1982), 129–133.

[K4] Krieg, A., Modular forms on half-spaces of quaternions, *Lect. Notes in Math.* 1143, Springer-Verlag, Berlin, 1985.

[K5] Kumano-go, H., *Pseudo-Differential Operators,* MIT Press, Cambridge, MA, 1981 (English translation).

[L1] Leray, J., Analyse lagrangienne et mécanique, Séminaire au Collège de France (1976–77), Paris.

[LP] Lewis, J.E. and Parenti, C., Pseudodifferential operators of Mellin type, *Comm. Part. Diff. Equ.* 8 (1983), 477–544.

[L2] Loos, O., *Bounded Symmetric Domains and Jordan Pairs,* Univ. of California, Irvine, 1977.

[M1] Melrose, R.B., Transformation of boundary problems, *Acta Math.* 147 (1981), 149–236.

[MT] Melrose, R.B. and Taylor, M.E., Near peak scattering and the corrected Kirchhoff approximation for a convex obstacle, *Adv. Math.* 55 (1985), 242–315.

[N1] Narasimhan, R., *Several Complex Variables,* The Univ. of Chicago Press, Chicago, 1971.

[P1] Perelomov, A., *Generalized Coherent States and their Applications,* Springer-Verlag, Berlin, 1986.

[P2] Petersson, H., Max Koecher's work on Jordan algebras, in: *Jordan Algebras* (W. Kaup, K. McCrimmon, H. Petersson, Eds.), de Gruyter, Berlin, 1994.

[RS1] Rempel, S. and Schulze, B.W., *Asymptotics for Elliptic Mixed Boundary Problems (Pseudo-differential and Mellin Operators in Spaces with Conormal Singularities),* Akad.-Verlag, Berlin, 1989.

[RS2] Rempel, S. and Schulze, B.W., Complete Mellin and Green symbolic calculus in spaces with conormal asymptotics, *Ann. Global Anal. Geom.* 4 (1986), 137–224.

[RV] Rossi, H. and Vergne, M., Analytic continuation of the holomorphic discrete series of a semi-simple Lie group, *Acta Math.* 136 (1976), 1–59.

[S1] Schmid, W., Die Randwerte holomorpher Funktionen auf hermitesch symmetrischen Räumen, *Invent. Math.* 9 (1969), 61–80.

[SS] Schrohe, E. and Schulze, B.W., Boundary value problems in Boutet de Monvel's algebra for manifolds with conical singularities I, preprint, Max-Planck Institut, Bonn, 1994.

[S2] Schulze, B.W., *Pseudo-differential Operators on Manifolds with Singularities,* North-Holland, Amsterdam, 1991.

[S3] Seeley, R., Refinement of the functional calculus of Calderon and Zymund, *Koninkl. Ned. Akad. Wet. Proceedings, Ser. A* 68 (1965), 521–531.

[S4] Siegel, C.L., Úber die analytische Theorie der quadratischen Formen. *Ann. Math.* 36 (1935), 527–606.

[S5] Springer, T.A., *Jordan Algebras and Algebraic Groups,* Springer-Verlag, Berlin, 1973.

[T1] Treves, F., *Introduction to Pseudodifferential and Fourier Integral Operators,* (2 vols.), The Univ. Series in Math., Plenum Press, New York, 1980.

[T2] Taylor, M.E., *Pseudodifferential Operators,* Princeton Univ. Press, Princeton, 1981.

[U1] Unterberger, A., Encore des classes de symboles, *Sém. Goulaouic-Schwartz,* Ecole Polytechnique, Paris, 1977.

[U2] Unterberger, A., Décompositions spectrales approchées associées à une famille d'oscillateurs harmoniques, *C. R. Acad. Sci. Paris* 287 (1978), 783–785.

[U3] Unterberger, A., Oscillateur harmonique et opérateurs pseudo-différentiels, *Ann. Inst. Fourier* 29 (1979), 201–221.

[U4] Unterberger, A., Quantification de certains espaces hermitiens symétriques, *Sém. Goulaouic-Schwartz* 1979–80, Ecole Polytechnique, Paris, 1980.

[U5] Unterberger, A., The calculus of pseudodifferential operators of Fuchs type, *Comm. Part. Diff. Equ.* 9 (1984), 1179–1236.

[U6] Unterberger, A., *Analyse Harmonique et Analyse Pseudo-différentielle du Cône de Lumière, Astérisque #156,* Soc. Math. France, Paris, 1987.

[U7] Unterberger, A., Quantification relativiste, *Mém. Soc. Math. France* 44–45, Paris, 1991.

[U8] Unterberger, A., Opérateurs pseudo-différentiels et analyse harmonique non commutative, Proc. Conf. *Equations aux dérivées partielles St.-Jean-de-Monts 1989,* Soc. Math. France, Paris, 1989.

[U9] Unterberger, A. and Unterberger, J., La série discrète de SL(2,ℝ) et les opérateurs pseudo-différentiels sur une demi-droite, *Ann. Sci. Ec. Norm. Sup.* 17 (1984), 83–116.

[U10] Unterberger, A. and Unterberger, J., A quantization of the Cartan domain BD I(q = 2) and operators on the light cone, *J. Funct. Anal.* 72 (1987), 279–319.

[U11] Unterberger, A. and Unterberger, J., Quantification et analyse pseudo-différentielle, *Ann. Sci. Ec. Norm. Sup.* 21 (1988), 133–158.

[U12] Upmeier, H., *Symmetric Banach Manifolds and Jordan C*-Algebras,* North-Holland Math. Studies #104, Amsterdam, 1985.

[U13] Upmeier, H., *Jordan Algebras in Analysis, Operator Theory, and Quantum Mechanics,* CBMS Regional Conf. in Math. #67, Amer. Math. Soc., Princeton, 1986.

[U14] Upmeier, H., Jordan algebras and harmonic analysis on symmetric spaces, *Amer. J. Math.* 108 (1986), 1–25.

[U15] Upmeier, H., Toeplitz operators on bounded symmetric domains, *Trans. Amer. Math. Soc.* 180 (1983), 221–237.

[U16] Upmeier, H., Weyl quantization of symmetric spaces: hyperbolic matrix domains, *J. Funct. Anal.* 96 (1991), 297–330.

[UU] Unterberger, A. and Upmeier, H., The Berezin transform and invariant differential operators, *Comm. Math. Phys.* 317 (1994), 1–35.

[W1] Warner, G., *Harmonic Analysis on Semi-Simple Lie Groups I,* Springer-Verlag, Berlin, 1972.

[W2] Weil, A., Sur certains groupes d'opérateurs unitaires, *Acta Math.* 111 (1964), 143–211.

[W3] Weyl, H., *Gruppentheorie und Quantenmechanik,* Hirzel, Leipzig, 1928.

Index of Notations

To define certain typographical symbols it has been necessary to assign letters to dummy variables; of course, other values can be met too (e.g., only P_x has been listed, not P_y); numbers refer to pages.

Special Symbols

Index